# GLOBAL ENVIRONMENT PROTECTION STRATEGY THROUGH THERMAL ENGINEERING

# GLOBAL ENVIRONMENT PROTECTION STRATEGY THROUGH THERMAL ENGINEERING

*Edited by*

**Keizo Hatta**
University of Tokyo
Hongo, Bunkyo-ku, Tokyo, Japan

**Yasuo Mori**
Tokyo Institute of Technology
Ohokayama, Meguro-ku, Tokyo, Japan

**OHEMISPHERE PUBLISHING CORPORATION**
A member of the Taylor & Francis Group

New York    Washington    Philadelphia    London

**GLOBAL ENVIRONMENT PROTECTION STRATEGY THROUGH THERMAL ENGINEERING**

Copyright © 1992 by Hemisphere Publishing Corporation. All rights reserved. Printed in the United States of America. Except as permitted under the United States Copyright Act of 1976, no part of this publication may be reproduced or distributed in any form or by any means, or stored in a database or retrieval system, without the prior written permission of the publisher.

1 2 3 4 5 6 7 8 9 0    BRBR    9 8 7 6 5 4 3 2 1

Cover design by Kathleen Ernst.
A CIP catalog record for this book is available from the British Library.

**Library of Congress Cataloging-in-Publication Data**

Global environment protection strategy through thermal engineering
  / edited by Keizo Hatta, Yasuo Mori.
      p. cm.
   "Papers from the final report of the research subcommittee in the
thermal engineering division of the Japan Society of Mechanical
Engineers . . . . "Energy Consumption and Thermal Engineering Study of
Global Environmental Problems" — Pref.
   Includes bibliographical references and index.

   1. Environmental protection — Congresses.  2. Energy conservation —
Congresses.   I. Hatta, Keizo.   II. Mori, Yasuo, date.
III. Nihon Kikai Gakkai.
TD169. G57       1992
628.5 — dc20                                                                 91-19345
ISBN 1-56032-146-6                                                             CIP

# Contents

Preface     vii

1    Introduction
      Y. Mori     1

2    Kinds of Thermal Energy and Present Conditions of Energy Consumption
      N. Nishikawa     8

3    Global Trends of Greenhouse Gases and Stratospheric Ozone
      H. Akimoto     22

4    Carbon Dioxide Problems

    4.1    Countermeasures to the Carbon Dioxide Problem in Hydrocarbon-Fired Plants
          Y. Mori     40

    4.2    New Combustion Technology for Carbon Extraction — Mechanism of Soot Precursor Formation
          K. Kuratani     63

    4.3    A New Approach to Reduce $CO_2$ Emission by Radiation-Controlled Combustion and Carbon Recovery
          R. Echigo     69

5    Hydrogen Fuel

    5.1    Production, Storage, Transportation and Utilization of Hydrogen
          E. Akiba     82

    5.2    Hydrogen Fueled Engines for Vehicle
          S. Furuhama     104

6    CFC Problems and Alternatives

    6.1    Regulations of CFC and Alternative Technologies
          A. Yabe     145

    6.2    CFC Alternatives
          T. Akiya     160

    6.3    Destruction Technology of CFCs
          T. Sugeta     167

7 NOx, SOx and Acid Rain Problems
  *T. Sano* **175**

8 Research and Development of Alternative Energy

  **8.1** Present Status of Development of Alternative Energy Technology from Environment Protection Point of View
  *T. Tanaka* **197**

  **8.2** Current Status and Future Prospects of Solar Energy Development
  *T. Tanaka* **200**

  **8.3** Current Status and Future Prospects of Geothermal Energy Development
  *Y. Yamada* **213**

9 High-Temperature Gas-Cooled Reactor and Its Application to Hydrogen Production
  *Y. Miyamoto* **234**

10 Advanced Energy Conservation

  **10.1** Energy-Saving Technologies in the Electric Power Industries in Japan
  *Y. Hara* **256**

  **10.2** Development of "Super Heat Pump" and Its Contribution to Global Environment Protection
  *A. Yabe* **282**

  **10.3** Cogeneration—Its Effects and Problems
  *A. Yabe* **305**

  **10.4** Highly Developed Energy Utilization by Use of Chemical Heat Pump
  *M. Hasatani* **313**

11 Efficiency Improvement of Energy-Related Facilities from an Environmental Viewpoint
  *K. Hijikata* **325**

Index **341**

# Preface

This book contains papers from the final report of the research subcommittee in the thermal engineering division of the Japan Society of Mechanical Engineers. The subcommittee was named "Energy consumption and Thermal Engineering Study of Global Environmental Problems."

In the past couple of years, various international conferences have been held by the United Nations Environmental Programme, ministries representing environmental agencies, and several other organizations. The increase of atmospheric carbon dioxide, the destruction of the ozone layers above the two poles, and acid rain have been primarily discussed, but few concrete countermeasures to these problems have been proposed. On the other hand, there may exist a comparatively wide gap between geoscientific discussion and the actual deterioration of the global environment.

Due to the difficulty of determining nationwide countermeasures by governmental committees and to the likelihood of unmanageable or uncooperative inter-governmental committees in the face of deteriorating environmental conditions, the subcommittee started its activities to develop concrete countermeasures for protecting the global environment. The subcommittee had several advantages such as not being influenced by external pressures and interferences, so that it could proceed with discussion from a purely academic and technical standpoint. The subcommittee consisted of researchers in charge of their respective areas of expertise concerning substantial environmental problems. The subcommittee members have made a concerted effort to report their research results and get them published, in contrast with international or national committees on environmental problems, which have served mainly as commentators and general reviewers. This book explains and proposes a strategy which includes the countermeasures and methodologies that the subcommittee obtained.

The editors would like to thank the contributors to this book, many of whom were subcommittee members. It should be noted that while the papers presented here were submitted to the subcommittee for review, each author is responsible for the content of his chapter. Any questions or comments should be addressed to the respective author, or contact Y. Mori at 5-9-8 Seijo, Setagaya-ku, Tokyo 157, Japan.

*Keizo Hatta*
*Yasuo Mori*

# 1. INTRODUCTION

**Yasuo Mori**
Tokyo Institute of Technology
Ohokayama, Megoro-ku, Tokyo Japan

At the time of the first and second energy crises, the great difference in energy situations between individual countries resulted in each country coping with energy problems in its own way. In Japan, for instance, the measures taken in relation to the problem of energy supply drastically changed both the capacity and the production processes in private sectors through the results of research and development and through the application of energy technology appropriate to each industry. Largely due to the enormous contribution by the private sector, Japan's amount of imported oil has not only been controlled, but decreased. This fact and the drop in oil prices helped to enhance economic and industrial activities, and the steady advances in technology in the past ten years or so have enabled Japan to become one of the major industrialized countries. However, since about 1988, global environmental problems such as the increase in carbon dioxide concentration in the atmosphere, the destruction of the ozone layer by chloro-fluorocarbons (CFC) and other gases, acid rain and the widening of dry areas such as deserts have been a matter of concern to many people in the world.

These global environmental problems are expected to become more and more serious in the years to come. Environmental problems have aspects quite different from energy problems and should be discussed not as a domestic affair for each country, but on a global scale and from an international standpoint. It is suspected, however, that each nation has its own specific environmental situation. As a result, although prompt solutions for environmental problems are required, few countermeasures have been reported so far. Under these circumstances, in 1988 the International Conference on Global Environment by the United Nations Environment Programme (UNEP) was held in Toronto, Canada, and an international conference on the environment was held by ministers from leading nations in The Hague, Netherlands. Many of the present global environmental problems are more or less related to thermal engineering. In consideration of the aspects of the relationship between global environment and thermal engineering and a lack of countermeasures to the global environmental problems known so far, a research subcommittee entitled "Energy Consumption and Thermal Engineering Study of Global Environmental Problems" was established by the Japan Society of Mechanical Engineers (JSME) in 1988. Most of this paper is compiled from the subcommittee's final report, originally written in Japanese and submitted to JSME Headquarters near the end of 1990.

The subcommittee worked on a strategy concerning the global environment, specifically focussing on concrete and systematical countermeasures to global environmental problems . With regard to some of the problems, the correlation, for example, between atmospheric carbon dioxide concentration and global warming has not necessarily been ascertained, though the increase in the concentration of carbon dioxide in atmosphere is clear enough. In international meetings, discussions on international policies and cooperation treaties on the carbon dioxide problem were made, but no countermeasures or methodology for solving the environmental problems have been reported. In other words, little or no progress has been made in this direction. It is cited, for example, that the amount of carbon dioxide sent out in the air is 20 billion tonnes annually, which might give us an impression of an extremely large amount, but it may be correct to add that the amount is only one hundred millionth of the total weight of the seawater. Another example is a report from an economic point of view. It predicts that the reduction in the discharge of carbon dioxide by 0.1% will keep the economical growth at about 1% which is not so desirable considering the recent condition of the world economy. In many industrialized countries, several environmentally-oriented committees have been organized with budgets for planning effective social policies or to initiate appropriate research projects for environmental protection. Supposedly, however, due to the rigid line drawn between the authorities of ministries, few nationwide projects or countermeasures to cope with environmental problems are reported. In sharp contrast to the fact that so far international or national committees have primarily consisted of commentators or general reviewers, the subcommittee members reported on their recent study and the results of their research. Thus, this book contains most of these results, although some of them are unpublished. In the subcommittee almost all of the problems associated with factors and substances related to the global environment were discussed. But, in consideration of the fact that the subcommittee's work was from the thermal engineering viewpoint, the following problems were not discussed in detail even though they were considered closely related to the global environment.

(a) The expected increase in world population and in global energy consumption.
(b) Utilization of the conventional water-cooled nuclear reactor as an alternative energy source.
(c) Protection of vastly stretching woods and forests.

Among the three items given above, problems (a) and (c) are mainly associated with underdeveloped or developing countries and involve many domestic concerns. Furthermore, they are not directly related to thermal engineering. Since the first energy crisis, the water-cooled nuclear reactor has played a leading role among alternative energies, and its electricity supply rate to the total supply has increased year by year. Even though the safety of the nuclear reactor is a matter of concern and has been fully investigated, several serious accidents occurred in the world in the past ten years. Since then, in most industrialized countries, careful operation and full training of reactor operators have been performed to prevent accidents. Due to this recent state of things, several projects to develop improved nuclear reactors of high safety and low power density, such as to have a surplus potential, are now carried on. An ABWR of boiling type and an APWR of pressurized type are now under construction . With the successful completion and operation of these nuclear plants, the future development of advanced and improved nuclear power plants will be possible. Under these circumstances, in this book, little discussion is made on them. On the other hand, the high temperature gas-cooled nuclear reactor is considered to produce hydrogen in addition to electricity with high thermal efficiency and safety. The R & D project of HTGR is now under thorough development mainly in West Germany and Japan. Because of a possible contribution of HTGR to the future global environment, its present status is described in Chapter 9.

Before giving an outline of every chapter, it should be noted that hydraulic power generation does not make a big contribution to the total world energy supply due to the difficulty of finding feasible locations from an environmental standpoint. The present major energy resource other than nuclear energy is hydrocarbon fuel, which has its own set of environmental problems. In other words, if a thermal plant burning a hydrocarbon fuel has a high thermal efficiency, it is not necessarily acceptable from an environmental standpoint if it sends out too much carbon dioxide and atmospheric pollutants.

To begin with, in Chapter 2, it is emphasized that, after the second energy crisis, electric power began to be more widely and conveniently used not only in modern industries but in home facilities. The ratio of consumed electric power to the total energy supply has remarkably increased. However, in response to the increasing consumption of electric power and in consideration of the recent difficulty constructing nuclear power plants timely, a fairly large amount of energy supply including electric power has come to be generated by burning hydrocarbon fuel in steam power plants and others. In Chapter 2, taking account of these facts, the aspect of the present and near future energy situations, the rapid increase in energy demand and the important role of electric power are discussed, particularly emphasizing the importance of an energy system and citing the Japanese situation as an example.

Chapter 3 is based on the observed data and results of a geophysical calculation; and, among the main environmental problems, it explains as concretely as possible, the global warming by carbon dioxide and methane etc., and the ozone layer destruction by CFC. Also, based on the results of a geophysical model calculation, the future atmospheric condition resulting from the increase in carbon dioxide is predicted. The model takes into account the interaction between atmosphere and the sea, and energy equations include radiative heat transfer between the planet surface and the atmosphere. Results would depend greatly on the model used in the calculation. The report made by the Roman Club in 1971 predicted the rise in oil cost and other problems including an increase in carbon dioxide concentration in the atmosphere. Presumably owing to a rather incorrect model, the oil price was predicted to be over $30 per barrel in the 1980s. One of the reasons for the discrepancy between predicted and actual prices could be attributed to the energy conserving efforts made by many industrialized countries. Because of the recent development of supercomputers and analytical models, it is becoming possible to predict the future atmospheric temperature, carbon dioxide concentration and the elevation of the sea level with higher accuracy. Global environmental problems concerning atmospheric carbon dioxide should be discussed in relation to its absorption into the sea. The problem of dissolution of carbon dioxide into the sea may be divided into two main categories, one of which is physical dissolution at the sea surface and the other is chemical dissolution that occurs mainly under the surface. As a matter of course, the global warming by carbon dioxide is discussed from the standpoint of physics, while the destruction of the ozone layer and the problem of acid rain are addressed from the standpoint of chemistry. Measures to solve these problems are discussed, putting importance on both fields. The later Chapters 4 and 5 discuss measures to control the atmosphere when hydrocarbon fuel is burned. In Chapter 6, the present state of the international restriction of CFC is scientifically explained and new alternative materials are discussed. New ways to decompose CFC before discharging it into the atmosphere are described which make some of the CFC acceptable in thermal equipment.

Among global environmental problems, the problem of carbon dioxide is considered to be the most serious one, as it is closely connected to life on the planet, not only human but animal and plant as well. As is well known, over 80% of the energy consumed on the worldwide scale is produced by burning hydrocarbon fuel. If a less expensive and environmentally acceptable process to recycle hydrogen is established in the future such as the thermochemical decomposition of water, and the hydrogen thus obtained is used as fuel with less generation of NOx, hydrogen will be evaluated as a future fuel. Lately, methanol is often recommended as an alternative to hydrogen and its production from sugar cane etc. by use of bioengineering technology, has been discussed recently. Careful research should be made not to let carbon dioxide escape in the production process of methanol. Carbon dioxide is absorbed in a plant by a photosynthetic reaction. Carbon absorbed by the reaction remains in the plant until the plant is withered. A part of the carbon stored in the plant may be changed to methane or methanol by a bioengineering technology, but the carbon remaining in the plant will react with oxygen in the atmosphere to produce carbon dioxide after the withering of the plant. Thus these processes, including photosynthesis and methane production from sugar cane, make no contribution to solve the carbon dioxide problem on the whole. A serious discussion was made in the subcommittee about problems created when methanol is used as automobile fuel. The conclusion on the use of methanol as an alternative fuel was pessimistic. Because of its low combustion heat and of an insuperable disadvantage of producing NOx and other toxic materials such as aldehyde, methanol is not considered to have so promising a future, even though it might be used as automobile fuel for the time being. Consequently, no further discussion is made about it in this book.

On the other hand, a high temperature, gas-cooled nuclear reactor (HTGR) pilot plant is operated by about 1100K helium in West Germany, and this technology is currently under research and development in both West Germany and Japan. It is considered to be the only energy system which generates not only electricity but produces hydrogen with the least environmental problems, and is discussed in detail in Chapter 9. In addition to HTGR and its applications, other hydrogen manufacturing technologies, its storage methods and registration laws which some countries have concerning handling hydrogen because of its flammability are discussed in Chapter 5. Research and development of reciprocating engines using hydrogen as fuel, in particular those engines using a spark-ignition plug similar to a gasoline engine, have been made in Japan, West Germany and other countries, and the recent results of their performances are explained in Chapter 9. As is well known, diesel engines with a large output of about several MW have been improved remarkably since the oil crisis so that they have a thermal efficiency of about 40% or more. Diesel engines are installed in many islands in the world for generating electricity, but exhaust gas from a diesel engine burning hydrocarbon fuel often produces a great deal of smoke containing such toxic substances as aldehyde. Therefore, for a couple of years, experiments have been performed with compression ignition reciprocating engines of diesel

## 2.1 Kinds of Thermal Energy and Present Conditions of Their Use

In view of the history of thermal energy consumption, from the discovery of fire several hundreds of thousands of years ago until the Industrial Revolution brought about by the invention of the steam engine in the latter half of the 1700s, almost oil thermal energy sources were wood or charcoal, with their intended use limited to cooking and heating.

Of course, resin and whale oil were also burned and used for illumination, but the scale of use was very small.

After the Industrial Revolution, the existing power sources such as waterwheels, windmills, human power and the animal strength of horses, oxen, etc., were rapidly replaced by the steam engine, a and coal came to be the daminant heat source.

Then, with the discovery of large oil fields in the Middle East after World War II, petroleum, cheaper and easier to treat, pushed coal out of the picture, and energy consumption drastically increased.

Reflecting back on the history of Japan the changeover to petroleum from coal, a domestically produced energy source which had been the mainstay of the nation's energy supply, began in the latter half of the 1950s. With this shift, the ratioof petroleum in the primary energy consumption pie graph increase rapidly, reaching, in fact, 78 % in 1973. However, after this, the two oil crises led to energy saving and introduction of alternative energy sources, resulting in the advent of the age of energy diversification.

At present, petroleum accounts for about 55 % of the primary energy supply, while natural gas and nuclear energy account for about 10 % each.

On the other hand, the share of coal, which was about 80 % in the early years, of Showa, has now been reduced to less tham 20 % for reasons of cost and environmental pollution.

TABLE 2.1.1 Proved recoverable reserves of fossil fuels in the world
(100 million tons in petroleum equivalent)

| Resources<br>Regions | Petroleum | Natural Gas | Coal | Total |
|---|---|---|---|---|
| China | 33 | 9 | 4471 | 4513 |
| Soviet Union and Eastern Europe | 82 | 390 | 1697 | 2169 |
| North America | 44 | 66 | 1406 | 1516 |
| Middle East | 898 | 313 | 1 | 1212 |
| Africa | 80 | 68 | 419 | 567 |
| Asia | 28 | 58 | 438 | 524 |
| Australia and New Zealand | 2 | 5 | 467 | 474 |
| Western Europe | 25 | 49 | 344 | 418 |
| Latin America | 170 | 60 | 104 | 334 |
| Total | 1362 | 1018 | 9347 | 11727 |
| Component Ratio (%) | 11.6 | 8.7 | 79.7 | 100 |

OGJ, issued at the end of 1989
World Energy Conference, 1989

Neverthless, Japan still imports some 80 % or more of its primary energy needs and almost 100 % of its petroleum requirements. Specifically Japan depends on the Middle East for about 80 % of the nation's oil imports [2.1.1.]. As such, the energy supply base of Japan is much more vulnerable than that of other developedcountries, and it will be necessary for Japan in the future to increase the use of nuclear power and expand the consumption of coal which, in comparison with petroleum and natural gas, has an overwhelmingly greater ratio of worldwide deposits of some 80 % among all fossil fuels. Moreover, it is also necessary to advance the development of hydropower and geothermal heat, twoindigenous energy sources, by taking into consideration compatibility with the environment. Indeed, the future of Japan, and the world, is closely tied to the realization of the use of natural energies such as solar radiation, biomass, geothermal, wind power, OTEC, etc.

Literature
[2.1.1 ] Ministry of International Trade and Industry. "Energy 87", Denryoku Shinposha, 1987.

## 2.2 Trend of Energy Demand

Looking back on the worldwide trends in energy consumption from the second oil crisis in 1979 to the present, the consumed amount of primary energy, which had been reduced following the oil crisis, turned sharply upward due to the lowering of oil prices, reaching about 8.0 billion tons in petroleum equivalent in 1988. OECD countries accounted for 50 % of this amount, with the remainder attributed to East European countries and the developing countries. It is interesting to note that the energy consumption amount of OECD countries had temporarily dropped and then returned in 1988 to the level of 1979, and that the energy consumption amount of the East European countries and the developing countries has been continuously increasing without a leveling off. Southeast Asian countries show a tendency of remarkable increase in energy consumption, particularly in recent years, with an increase of 10 % or more in 1988, while OECD countries, show an increase rate of about 3 % for the same year.

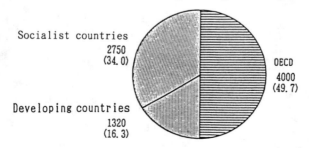

FIGURE 2.2.1 Consumption amount of primary energy in the world (1988)
(Source: BP Statistics, 1989)

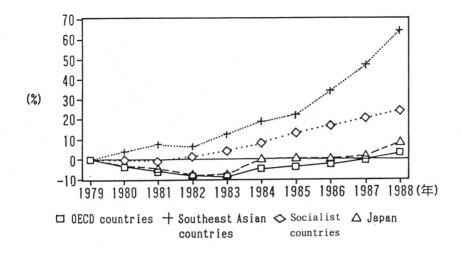

FIGURE 2.2.2 Increase rates of energy consumption, taking 1979 as the base (Source: BP Statistics, 1989)

In 1984 the energy consumption of Japan marked a substantial increase of 8 % over the preceding year, followed by an increase rate of about 0.6 % annually until 1987, which is considerably lower than that of 1.7 % on the average for OECD countries.
However, in 1988 Japan's consumption rate increased 6.2 % over the previous year, which was the highest growth rate recorded by the OECD countries. The background for this radical increase lies in the following factors: stabilization of oilprices at the bottom-most level; reduction in energy costs due to the higherexchange rate of the yen; high economic growth; and structural changes in industries and people's lifestyles.
Viewed from the breakdown of energy end use, weight of industry has been reduced while the sectors of household and transportation show high growth. By source of energy, electricity shows a remarkable increase, particularly in the household sector. As factors of the increase in electricity demand, we can point to the increasing demand for heat-pump airconditioners in households, reflecting the public's constantly growing awareness of amenity and the demand increases for robots and high-tech equipment in industry.

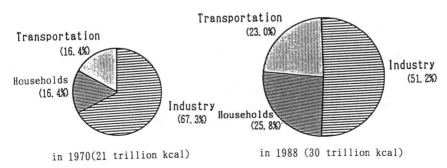

FIGURE 2.2.3 Ratios of energy end use by sector in Japan (1988)

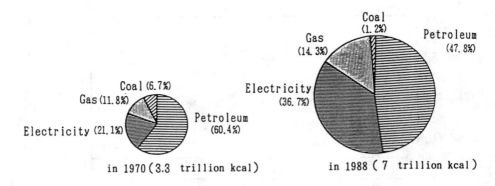

FIGURE 2.2.4  Consumption ratios by energy source in household sector

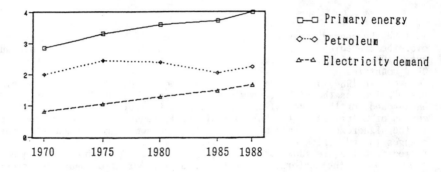

FIGURE 2.2.5  Growth of electricity demand in Japan

(Source: BP Statistics in 1989 and others)

2.3 Energy Consumption and $CO_2$ Emission Amount In Japan

Chartacteristics of Japan in the demand structure of primary energy is that the demand ratio of petroleum is greater and that of coal is smaller than the world level. The unit amount of $CO_2$ emission from fossil fuels is greatest for coal. If we take the $CO_2$ emission amount from coal to be 100, then that from petroleum would be 80 and that from natural gas, 60. Nulear power plants generate no $CO_2$. Accordingly, $CO_2$ ession amount per energy consumption in Japan is lower than that of other countries.

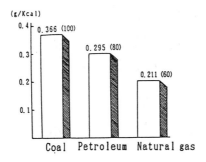

FIGURE 2.3.1 Unit of $CO_2$ emission amount from various fossil fuels

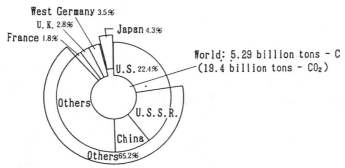

FIGURE 2.3.2 $CO_2$ emission amount from fossil fuels (1985)
Source:Literature [2.3.1]

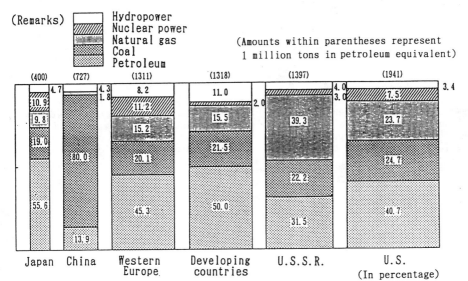

FIGURE 2.3.3 Component ratios of primary energy supply sources
(Source: BP Statistics, 1989)

This is particularly true for the fuel structure of electricity. Owing to the save oil policy after the oil crises, the electric power industry in Japan has introduced LNG and nuclear power to advance the diversification of its fuel portfolio, and its dependence on petroleum has now been reduced to 30 % or less from 71 % at the time of the first oil crisis.
The ratio of coal consumption is also quite low in comparison with other countries.

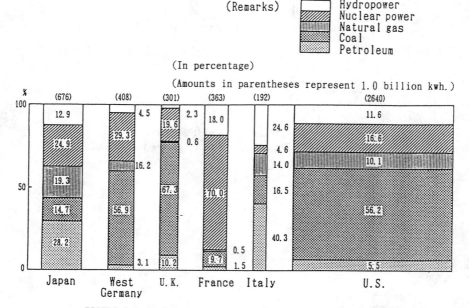

FIGURE 2.3.4   Component ratios of power generation sources in advanced countries  (1986)
Source: Literature [2.5.2]

Since the component ratios of petroleum and coal are small in this way, $CO_2$ emission amount per generated electricity in Japan is the second lowest among the advanced countries. The reason why France has the smallest emission amount, about 0.02 t per 1,000 kwh, is that its dependence on nuclear power is 70 %, an extremely higher figure than that in other countries.

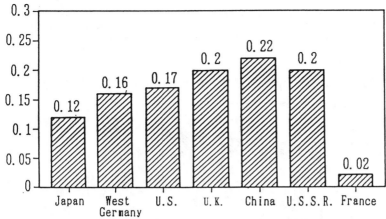

FIGURE 2.3.5  CO₂ emission amount per generated electricity

The unit of $CO_2$ emission amount per energy consumption in the sector of electricity generation is lower than that in the sectors of industries and transportation.

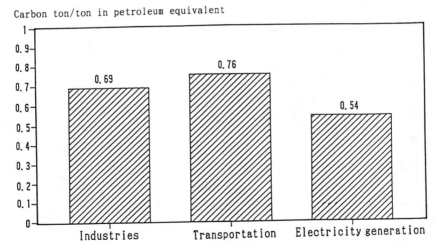

FIGURE 2.3.6  Unit of CO₂ emission amount by sector in Japan

Literature
[2.3.1] Revised by Jiro Kondo, [Verification] Energy and Abnormalities of the Earth, Energy Journal, 1989.

## 2.4 $CO_2$ reduction effects through improvement of energy use efficiency

As previously discussed, through promotion of oil alternative energy sources following the two oil crises, that is, the shift to LNG and nuclear power, the $CO_2$ generation amount per energy consumption in Japan has been reduced.

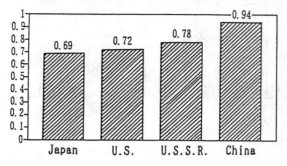

FIGURE 2.4.1 Comparison of $CO_2$ generation amounts per energy consumption

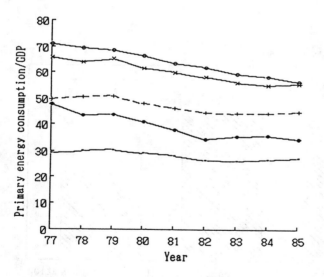

FIGURE 2.4.2 Unit of primary energy supply per GDP (10 thousand tons in petroleum equivalent/billion US dollars of 1980)

OECD:NATIONAL ACCOUNTS 1960 ~1987
U.N. statistics 1987
DOE :Monthly Energy Review 1987
British Energy Department Statistics 1986
VIK :Statistik der Energiewirtshaft 1986/87
Comprehensive Energy Handbook 1987~1988
EDF :ANNUAIRE STATISTIQUE 1987

## 2.6 Global Measures for Reducing $CO_2$

A trial calculation [2.6.1] estimates that a 20 % reduction in the amount of $CO_2$ emission in Japan may be achieved, but with much difficulty, by restricting the growth of energy demand to 0.5 % /year, the same level as in the 1980s following the oil crises, by shifting the present use of coal in thermal power plants wholly to LNG, and by removing and disposing of 50 % of the $CO_2$ generated in thermal power plants into the sea or elsewhere. This is a far from easy task.

As for the technology to remove $CO_2$ generated in thermal power plants alone, there are many problems to be solved in conducting this research. Also, even if removing $CO_2$ is successful, the problem of how to dispose of tremendous amount of $CO_2$ in formidable. Of course, studies will be advanced energetically in the future, but, at this stage, the technology is extremely difficult. On the other hand, thefact that Japan is a country with high energy efficiency means that we there is little room for efficiency improvement. That is to say, when the target level of $CO_2$ reduction is determined in the future, Japan may be forced to cope with far severer conditions than other countries depending on the contents of this decision. As the overall energy demand, especially the demand for electric power is growth increasingly, it is prerequisite for Japan to overcome the trilemma of economic growth, increases demand and $CO_2$ reduction. Thus, it is necessary to exert strenous efforts, beginning with what we can do now, for technological development aimed at the most suitable mix of energy, further improvement of energy efficiency and $CO_2$ reduction.

As a cause of $CO_2$ increases, apart from the combustion of fossil fuels, one can refer to large-scale destruction of our forests. The amount of $CO_2$ increase due to forest destruction is reported to be 1.0~2.0 billion tons carbon/year, a mind-boggling amount equaling some 40 % of $CO_2$ generated from the combustion of fossil fuels all over the world. Accordingly, the prevention of deforestation is an effective measure.

Another helpful means is recycling of used paper in developed countries. This important measure can be realized at once and will lead to the protection of forests.

Literature
[2.6.1] Edited by Workshop on Mankind and Energy, Global Environment and Mankind, Center for Energy Saving (Foundation), 1989.
[2.6.2] Masayuki Tanaka, Wormikng up of the Earth, Yomiuri Science Selection 23, 1989.

Conclusion
As described herein, countermeasures against warming up of the earth are extremely difficult to realize. Nevertheless, We can not stand idly by and do nothing. Together with making efforts towards the most suitable mix of energy and improvement of energy efficiency, it is necessary to devise countermeasures from a long-term viewpoint, beginning with what we can do immediately and including intensive technological efforts to achieve direct measures for reducing $CO_2$ emissions. The electric power industry is not only putting foward the introduction of nuclear power but also actively promoting technological development for improvement of efficiency in energy use and for utilization of natural energy. At the same time, the industry has begun to examine these measure carefully so as to directly remove and dispose of $CO_2$ from exhaust gases of power plants.

It can be said that the conception of "Think Globally, Act Locally" is required for the solution of $CO_2$ problems.

# 3. GLOBAL TRENDS OF GREENHOUSE GASES AND STRATOSPHERIC OZONE

**Hajime Akimoto**
National Institute for Environmental Studies
Onogawa, Tsukuba, Ibaraki 305, Japan

**Global atmospheric environment system**

The earth is a closed system in which atmosphere, hydrosphere and biosphere are inter-related by exchanging energy and chemical species. Mankind in itself is a member of biosphere, and is to be harmonized with the earth system. Accompanying the increase of population and energy consumption after the industrial revolution, however, the impact of human activities to the system exceeded the extent of the expected harmonization, which has resulted the global environmental pollution. Figure 3.1 schematizes important elements of global atmospheric environmental issues such as related to greenhouse threat, stratospheric ozone destruction, and acid deposition. The structure of the global atmospheric environment system perturbed by the impact of human activities would be summarized as follows referring to Fig. 3.1.

(1) Global change of background concentrations of atmospheric trace gases directly emitted by human activities such as $CO_2$, $CH_4$ chlorofluorocarbons (CFC's), CO, $NO_x$, $SO_2$, etc.

(2) (a) Physical perturbation to the earth system such as the enhanced greenhouse effect accompanying the increase of earth surface average temperature, climate change, rise of sea level, etc. due to the increase of these trace gases.

(b) Chemical perturbation to the earth system such as the effect to the stratospheric chemistry resulting the ozone destruction, and the effect to tropospheric chemistry resulting the tropospheric ozone increase and enhanced acid deposition.

(c) Feedback between the phenomena caused by the physical and chemical perturbations, e. g. the influence of greenhouse effect (stratospheric temperature decrease) to the stratospheric ozone destruction, the effect of the increase of tropospheric ozone (greenhouse gas) to the global warming, increase of UV radiation at the ground surface due to the stratospheric ozone destruction, which would result further increase of tropospheric ozone, etc.

(3) The effects of physical and chemical environment change on the terrestrial and aquatic ecosystems, and their feedback to the earth system through the change of biogeochemical cycles.

(4) Social and economical effects of global physical and chemical change, and their feedback to human activity through the development of new technology and social system.

This chapter concerns recent topics on the global trend of the concentrations of selected greenhouse gases ($CO_2$, $CH_4$, $N_2O$ and CFC's) and stratospheric ozone.

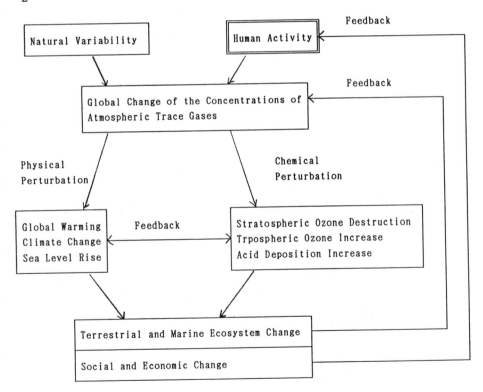

FIGURE 3.1 Global atmospheric environment system.

## Carbon dioxide ($CO_2$)

Analysis of air bubbles in ice core samples provides us information on the variation of the concentration of $CO_2$ in the paleo-atmosphere of the earth. From the analysis of the 2000m deep core sample drilled at the Vostok base in the Antarctica, it has been revealed that there is a high positive correlation between the atmospheric $CO_2$ concentration and the global surface temperature [3.1]. Thus, the $CO_2$ concentration was less than 200 ppm in the glacial periods(20 kyr and 150 kyr ago), and about 280 ppm in the interglacial(120-130kyr ago) and the post-glacial period(recent 10 kyr), while the global surface temperature was varied about 5°C between these periods. More recent trend of atmospheric concentration of $CO_2$ has also been obtained from ice core analyses, which shows that the $CO_2$ level had been constant within the range of 275-280 ppm during the last few thousand years until the increase of $CO_2$ started just in the middle of the 18th century [3.2].

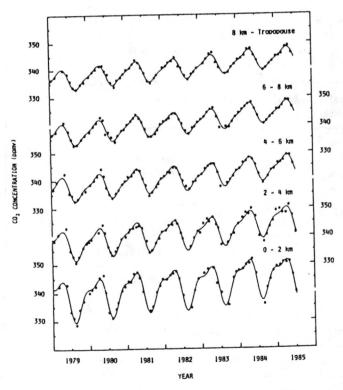

FIGURE 3.2 Change of the $CO_2$ concentration at the different altitude in the troposphere over Japan [3.3].

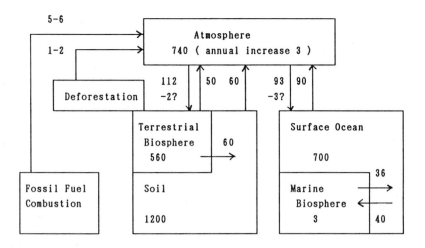

FIGURE 3.3 Carbon cycle. Units are Pg($10^{15}$g) for reservoir contents and Pg/yr for transfer rates.

Figure 3.2 depicts the data of Tanaka et al. [3.3] showing the increase of $CO_2$ concentration at the different altitude in the troposphere over Japan in the last ten years. The concentration increased from 336.7 to 350.3 ppm during 1979 and 1988; the average increase rate is thus 1.5 ppm/yr(0.43%/yr). It can be noted that the phase of the seasonal cycle delays at the higher altitude, which verifies that the source of $CO_2$ is on the ground. The seasonal cycle showing the maximum in winter and minimum in summer, apparently reflects the activity of plants.

In order to discuss the mechanism of $CO_2$ increase in the atmosphere, the budget of carbon which cycles between the atmosphere, hydrosphere and biosphere in the from of $CO_2$, carbonate, and organic carbon, has to be considered. Figure 3.3 shows the reservoir contents and transfer rates of carbon between the reservoirs. Since 1 ppm of $CO_2$ in the atmosphere corresponds to 2.12 PgC(1 petagram = $10^{15}$g, which is equivalent to 1 Gt) of total carbon in the atmosphere, 280 and 350 ppm of $CO_2$ before the industrial revolution and at present amount to 594 and 740 PgC of total atmospheric carbon, respectively. Emission of $CO_2$ from fossil fuel combustion(including emission from cement industry) has increased approximately exponentially(about 4%/yr) except the period of World War I and II, and the recession in 1930's. In recent years, the increase rate decreased to 2 %/yr after the oil crises of 1973, and then the emission rate has been nearly constant at 5.3 PgC/yr after 1979, but tends to increase again after 1985 [3.4]. Annual emission rate of $CO_2$ from fossil fuel burning is estimated as 5.7 PgC at present, and the accumulated emission during 1850-1986 as 200 $\pm$ 20 PgC.

Another source of $CO_2$ is deforestation, which is mostly in the tropical region in recent years. The emission rate due to deforestation has been estimated to be 1.6 $\pm$ 0.8 PgC/yr [3.5]. Model calculation of latitudinal distribution of $CO_2$, however, deduced the net $CO_2$ emission from the terrestrial biosphere to be 0 - 0.9 PgC/yr in the last 30 years [3.6], which is much lower than the above ecological estimate. The discrepancy might be due to the increase of primary production of plants accompanying the increase of atmospheric $CO_2$ concentration (fertilizing effect). On the contrary, deforestation in the 19th century and at the beginning of 20th century has been occurred in the temperate zone (maximum rate of $CO_2$ emission, 0.5 PgC/yr) and the accumulated emission of carbon due to the deforestation between 1850 and 1958 has been estimated to be 115 $\pm$ 35 PgC [3.7].

From these data total amount of carbon emitted as $CO_2$ during 1850 and 1987 sums to about 315 PgC which should be compared to the atmospheric increase of 144 PgC corresponding to the change of $CO_2$ concentration from 280 to 348 ppm. Thus, about 45 % of total carbon emitted remains in the atmosphere. In the last 28 years since 1958, the percentage of carbon remaining in the atmosphere is constant, 58 % of emission from the fossil fuel combustion, or about 40 % of the sum of the fossil fuel combustion (5.7 PgC/yr) and deforestation (1.6 PgC/yr). This would mean that the difference of 4.1 PgC/yr between the total emission of 7.3 PgC/yr and the atmospheric increase of 3.2 PgC/yr has to be removed from the atmosphere annually. Since the uptake of $CO_2$ by the ocean is estimated to be ca. 2 PgC/yr, the remaining ca. 2 PgC/yr has to be removed by a "missing sink". Fertilizing effect for the terrestrial plants could account for the discrepancy [3.8] as noted above.

### Methane ($CH_4$)

In addition to $CO_2$, atmospheric concentration of other greenhouse gases, $CH_4$, $N_2O$ and chlorofluorocabons(CFC's) are also known to be increasing in recent years. The concentrations of $CH_4$ in the paleo-atmosphere have also been obtained from the data of ice core analyses. As is the case of $CO_2$, the atmospheric concentration of $CH_4$ has a high correlation with temperature over the past 160 kyr; the concentration being less than 350 ppb in the glacial period and about 650 ppb in the interglacial period thus the concentration ratio between the glacial-interglacial episodes being close to a factor of two [3.9]. In the last 3000 yr, the concentration of $CH_4$ has been constant at 700 ± 100 ppb until they starts to increase in the 19th century [3.10]. Present level of $CH_4$ is 1.72 ppm and the increase rate is 16 ± 1 ppb/yr (0.9 ± 0.1%/yr) [3.11].

Lists of emission sources of $CH_4$ are wetlands, oceans, freshwater, and methane hydrate destabilization as natural sources, and rice paddies, cattle, biomass burning, landfills, natural gas drilling(including venting and transmission), and coal mining as anthropogenic sources. Figure 3.4 shows an estimate of annual emission of $CH_4$ from identified

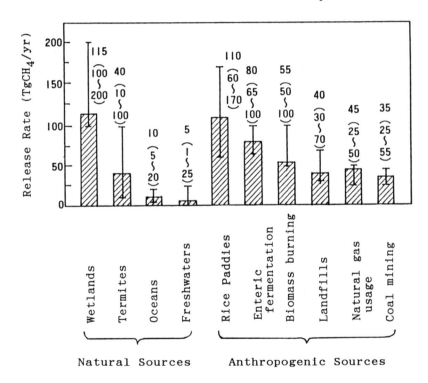

FIGURE 3.4 Global annual release rate of $CH_4$ from identified sources [3.12]. Enteric fermentation includes wild animals as well as cattles. Natural gas usage includes drilling, venting and transmission.

sources [3.12] . Rice paddies are recognized as one of major $CH_4$ sources. Recent study has revealed that emission strength of $CH_4$ depends on 1) season, 2) agricultural management style (types of fertilizer, water management, planting density of rice seedling, etc. ), and 3) nature of soils (types, acidity, oxidation-reduction potential, temperature, nutrients, etc. ) [3.13] . Rice production has doubled since 1940's, which possibly increased the emission of $CH_4$ from paddies. Biomass burning is another major anthropogenic source of $CH_4$. Biomass burning has been increasing in a global scale since the last century, which would have increased the emission of $CH_4$. More data on the emission factor, burning size of land and burning style is necessary to obtain more accurate estimate.

Table 3.1 gives an estimate of $CH_4$ emission strength from various sources in Japan [3.14] . As shown in Table 3.1., total emission of $CH_4$ in Japan has been estimated about 0.99 $TgCH_4$/yr which is about 0.18 % of the global emission. Among the sources, rice paddies and cattle, are the major contributor followed by waste treatment, particularly landfills. It should be noted that fossil fuel combustion is a minor contributor among anthropogenic sources. The estimate for the natural gas usage is based on an assumption that 2 % of the

Table 3.1 Estimate of annual methane release rates for identified sources in Japan

| Sources | Annual Release Rates $10^9$ $gCH_4$ |
|---|---|
| Anthropogenic Sources | |
| Rice paddies | 380 |
| Cattle | 280 |
| Landfills | 215 |
| Other waste treatments | 35 |
| Fossile fuel combustion | 15 |
| Venting, trnsmission gas drilling | 30 |
| Natural Sources | |
| Wetlands | 25 |
| Lakes | 10 |
| Total | 990 |

total production leaks to the atmosphere in the total processes of drilling, venting and transmission. Contribution of natural sources is also minor in Japan due to the limited area size of wetlands and lakes.

Methane emitted into the atmosphere is dissipated by the reaction with OH radicals in the troposphere,

$$CH_4 + OH \text{ --- } CH_3 + H_2O \qquad (3.1)$$

Average residence time of $CH_4$ in the atmosphere is thus determined by the loss rate of $CH_4$ by reaction (3.1). Using the global annual average OH concentration of $(7.7 \pm 1.4) \times 10^5$ molecule /$cm^3$, the atmospheric lifetime of $CH_4$ can be deduced to be $9.6 \pm 2$ yr [3.15] . Increase of anthropogenic emission of CO results in the decrease of OH concentration due to the competitive reaction,

$$CO + OH \text{ --- } CO_2 + H \qquad (3.2)$$

for the OH radicals, which would reduce the dissipation rate of $CH_4$ and bring the increase of global concentration of $CH_4$ as a result [3.16] .

Total amount of $CH_4$ in the atmosphere can be calculated to be 4800 $TgCH_4$ from the global distribution at present and the annual increase of $0.9 \pm 0.1$ % corresponds to the emission strength of $45 \pm 5$ $TgCH_4$ /yr. In order to stabilize the $CH_4$ concentration at the present level, 10 - 15 % reduction of anthropogenic emission of $CH_4$ should be required as depduced from the above emission strength divided by the total anthropogenic emission strength of 365 $TgCH_4$ /yr as given in Fig. 3.4

## Nitrous oxide ($N_2O$)

The concentration of $N_2O$ has increased from $285 \pm 5$ ppb before the 19th century to 310 ppb at 1987 with the recent increase rate of 0.2-0.3 %/yr [3.17, 3.18] .

Major natural sources of $N_2O$ are tropical and temperate forest soil, and ocean. Anthropogenic sources are fertilized soil(including freshwater), biomass burning, and fossil fuel combustion. Recent estimate of global $N_2O$ emission from identified sources is given in Figure 3.5. Compared to previous estimate, emission rates from biomass burning and fossil fuel combustion have been reduced by taking into account more recent data [3.19] .

Global total emission of $N_2O$ can also be assessed from

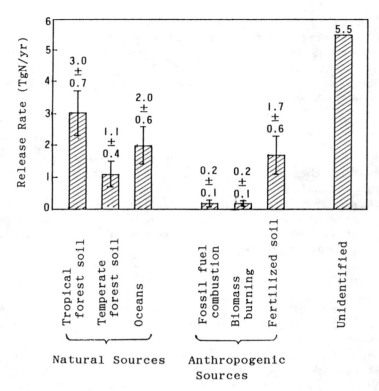

FIGURE 3.5 Global annual emission of $N_2O$ from identified sources [3.19].

the total loss rate in the atmosphere. Since $N_2O$ is thought to be stable in the troposphere, atmospheric lifetime of $N_2O$ is determined by the photolysis and the reaction with excited oxygen atoms, $O(^1D)$, in the stratosphere.

$$N_2O + h\nu \longrightarrow N_2 + O(^1D) \qquad (3.3)$$

$$N_2O + O(^1D) \longrightarrow 2NO \qquad (3.4)$$

Average lifetime of $N_2O$ due to these reactions has been estimated as $150 \pm 30$ yr by model calculations. Since the total mass of nitrogen in $N_2O$ in the atmosphere can be calculated to be about 1500 TgN, annual loss rate of $N_2O$ should be $10 \pm 3$ TgN/yr and the annual increase of $N_2O$ concentration, 0.2 - 0.3 % corresponds to the emission strength of 3.0 - 4.5 TgN/yr. Therefore, in order to balance the loss rate and annual increase of atmospheric concentration, global total emission rate of $N_2O$ should be 13.7 ± 3.5 TgN/yr. This is in contrast to the estimated emission rate of 8.2 ± 2.4 TgN/yr from identified natural and anthropogenic sources. The discrepancy of 5.5 TgN/yr must be ascribed to identified sources. Further, if we assume that the natural emission strength should correspond to the preindustrial $N_2O$ concentration of 280 ppb and has not been changed, it amounts to 9.4 ± 1.8 TgN/yr. The difference between the total global emission and the total natural emission of $N_2O$ given above, ca. 4.3 Tg/yr may be ascribed to the total anthropogenic emission. Thus, referring to Figure 3.5, out of 5.5 TgN/yr of unknown emission, ca. 3.3 TgN/yr and 2.2 TgN/yr would be ascribed to natural and anthropogenic sources, respectively, which are not specified at this stage. In order to stabilize the $N_2O$ concentration at the present level, reduction of more than 70 % of the anthropogenic emission is necessary from the above values.

## Chlorofluorocarbons(CFC's)

Chlorofluorocarbons are of importance both as ozone destructing gases and as greenhouse gases. Table 3.2 shows the tropospheric concentrations and annual increase rates of

Table 3.2 Tropospheric concentrations and annual increase rates of halocarbons (1987)

| Halocarbons | | Mixing Ratio ppt | Annual Increase Rates | | Atmospheric Lifetime yr |
|---|---|---|---|---|---|
| | | | pptv | % | |
| $CFCl_3$ | (CFC-11) | 240 | 9.5 | 4 | 75 |
| $CF_2Cl_2$ | (CFC-12) | 415 | 16.5 | 4 | 110 |
| $CF_3Cl$ | (CFC-13) | 5 | | | 400 |
| $C_2F_3Cl_3$ | (CFC-113) | 45 | 4-5 | 10 | 90 |
| $C_2F_4Cl_2$ | (CFC-114) | 15 | | | 180 |
| $C_2F_5Cl$ | (CFC-115) | 5 | | | 380 |
| $CCl_4$ | | 140 | 2.0 | 1.5 | 40 |
| $CHF_2Cl$ | (HCFC-22) | 100 | 7 | 7 | 20 |
| $CH_3Cl$ | | 600 | | | 1.5 |
| $CH_3CCl_3$ | | 150 | 6.0 | 4 | 7 |
| $CBrClF_2$ | (halon 1211) | 1.7 | 0.2 | 12 | 25 |
| $CBrF_3$ | (halon 1301) | 2.0 | 0.3 | 15 | 110 |
| $CH_3Br$ | | 10-15 | | | 1.5 |

CFC's as of 1987 [3.20]. Among the chemicals cited in Table 3.2., most of them are anthropogenic origin except methyl chloride ($CH_3Cl$) and methyl bromide ($CH_3Br$). Chlorofluorocarbons are used as aerosol propellant (CFC-11, 12, 114), refrigerant (CFC-11, 12, 114. HCFC-22), foaming agent (CFC-11, 12), cleaning solvent (CFC-113. $CH_3CCl_3$. $CCl_4$), etc., and significant increase of the atmospheric concentrations of these chemicals has been reported as shown in Table 3.2. Among the CFC's, the concentrations of CFC-12 is the highest followed by CFC-11; their increase rates are both ca. 4 %/yr. It should also be noted that the increase rates of CFC-113 which is used as cleaning agent for semiconductors, and HCFC-22 which is to be used as CFC alternative, are much higher, 10 %/yr and 7 %/yr, respectively, although their present concentration is lower than CFC-11 and 12. Except these CFC's and HCFC, 1,1,1-trichloroethane($CH_3CCl_3$) and carbon tetrachloride($CCl_4$) contribute significantly to atmospheric organic chlorine. Their concentrations exceeds 100 ppt and annual increase rates are 4 and 1.5 %, respectively. Halon-1211 and

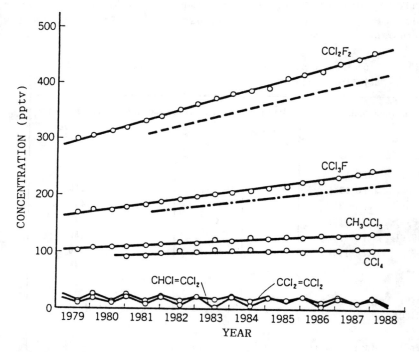

FIGURE 3.6 Increase of halocarbons at Hokkaido( 42-45° N, solid line) and Antarctic Syowa Base(broken and dash-dot line) [3.21].

1301 which are used as fire extinguisher, have been increasing at the rate of more than 10 %/yr.

Figure 3.6 shows the trend of the concentrations of selected halocarbons monitored at Hokkaido, northernmost part of Japan, and the antarctic Syowa base [3.21]. About 30 % discrepancy in concentrations of $CH_3CCl_3$ and $CCl_4$ as compared with the data in Table 3.2 has been ascribed to the calibration technique [3.22]. Anthropogenic CFC's are mostly emitted in the northern hemisphere and transported to southern hemisphere within 1-2 years. Therefore, the concentrations of long-lived CFC's at antarctic is about 10 % lower than those in the northern hemisphere mid-latitude, corresponding to the delay of 1-2 years.

**Stratospheric ozone**

It has been revealed that the stratospheric ozone over the antarctic has been destroyed drastically in spring time during the last decade. The total ozone in 1987 and 1989 was as low as 50 % of the original value in October and the destruction was found to occur in the lower stratosphere, 12 - 24 km above the ground as opposed to the "Molina-Rowland theory" of ozone destruction by chlorofluorocarbons, which predicts the loss of ozone in the upper stratosphere of around 35 - 40 km.

$$CFC + h\nu \longrightarrow Cl + \cdots \quad (3.5)$$

$$Cl + O_3 \longrightarrow ClO + O_2 \quad (3.6)$$

$$ClO + O \longrightarrow Cl + O_2 \quad (3.7)$$

The stratospheric ozone destruction commonly called "antarctic ozone hole", is thought to be caused by the conversion of rather stable chlorine compounds such as $ClONO_2$ and $HCl$ to the more active species such as $HOCl$ and $Cl_2$ by surface reactions on the polar stratospheric cloud(PSC). Simultaneous removal of $NO_2$ by a surface reaction is another important prerequisite for the ozone hole since the $NO_2$ scavenges $ClO$ radical, a main chain carrier of the chlorine-catalyzed ozone destruction. The main reaction sequence of "ozone hole theory" is as follows.

$$ClONO_2 + H_2O \xrightarrow{PSC} HOCl + HNO_3 \quad (3.8)$$

$$ClONO_2 + HCl \xrightarrow{PSC} Cl_2 + HNO_3 \qquad (3.9)$$

$$NO_2 + O_3 \longrightarrow NO_3 + O_2 \qquad (3.10)$$

$$NO_2 + NO_3 + M \xrightarrow{M} N_2O_5 + M \qquad (3.11)$$

$$N_2O_5 + H_2O \xrightarrow{PSC} 2HNO_3 \qquad (3.12)$$

$$HOCl + h\nu \longrightarrow OH + Cl \qquad (3.13)$$

$$Cl_2 + h\nu \longrightarrow 2Cl \qquad (3.14)$$

$$Cl + O_3 \longrightarrow ClO + O_2 \qquad (3.6)$$

$$2ClO \longrightarrow Cl + OClO \text{ etc.} \qquad (3.15)$$

The PSC is formed by the condensation of $H_2O$ and $HNO_3$ in the stratosphere or the temperature lower than $-80°C$ in the antarctic winter. After the irradiation of sunlight starts in August, ClO radical concentration is observed to be 50 - 100 times higher than normal value, and anti-correlation between the concentration of ClO and $O_3$ has been found when the ozone hole has developed [3.23]. These results strongly suggest that the antarctic ozone destruction is caused by chlorine-catalyzed reactions due to the increased level of CFC's.

Another recent finding is that the ClO concentration in the arctic stratosphere in January-February in 1989 was as high as in the antarctic, i. e. 50-100 times higher than normal, and ozone is destructed locally [3.20]. The much smaller scale destruction of ozone in the arctic region as compared to the antarctic is thought to be due to the difference in the meteorological condition, i. e. polar vortex in which the Cl-catalyzed ozone destruction proceeds with time scale of a month in the antarctic, disappears in the arctic mostly before the ozone destruction develops.

Several analyses of the long-term trend of total ozone other than the polar regions have been reported since the Ozone Trend

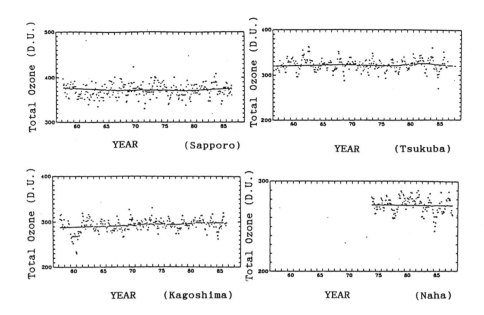

FIGURE 3.7 Trend analysis of total ozone around Japan [3.24].

Panel Report [3.20]. It should particularly be noted that the total ozone in the northern hemisphere midlatitude (30 - 64° N) has decreased 3.2 - 4.0 % during the past 17 years 1969 and 1988 according to the reports [3.20]. An analysis conducted in Japan using a method developed in this country resulted the conclusion that the total ozone decreased 2.1 % in winter during the period but no significant trend was found in summer [3.24]. Further, total ozone variation is not uniform in a latitudinal zone, and the trend is different at a different longitude. The decreasing trend is significant in Northern America and Europe but not apparent around Japan. Figure 3.7 shows the total ozone trend at the four Japanese stations at Sapporo, Tsukuba, Kagosima, and Naha over the last 25 years. As shown in the figure, no significant decrease is observed in these stations and even a slight increase is observed at Kagosima.

Information on the altitude dependence of ozone decreasing trend in the mid-latitude is of great importance from the mechanistic point of view. If the ozone destruction occurs by the Cl-catalyzed homogeneous reactions, the ozone decreasing trend should be found in the upper stratosphere, while if it occurs by another mechanism e. g. involving heterogeneous reactions

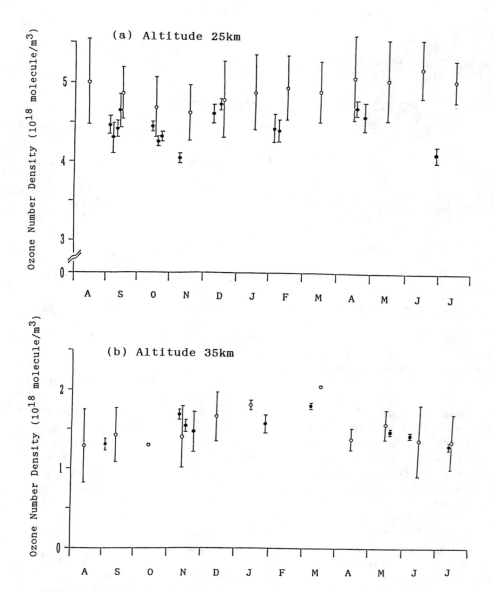

FIGURE 3.8 Comparison of monthly averaged values of ozone number density observed by NIES ozone lidar (⊢•⊣) in 1988-89 and the 20 years averaged values observed by ozone sonde (⊢○⊣) in 1968-87 at Tsukuba, Japan [3.24].

on aerosol, it may be pronounced in the lower stratosphere. Satellite data suggested the decreasing trend at both around 40 km and 20 km in the last 7 years in the mid -high latitude [3.20] . It is interesting to note that the decreasing trend in the lower stratosphere is consistent with the decrease of the total ozone described above. Figure 3.8 shows the comparison of ozone lidar measurement data of 1988-89 obtained at our Institute and the 20 years average of ozone sonde data at Tsukuba at 25 and 35 km in altitude. It can be noted that the ozone level during the lidar observation is about 10 % lower than the 20 years average only at 25 km and no difference from the average was observed at the higher levels. Such trend also been confirmed by ozone sonde data in the last two years. Although it is not clear at this stage whether these decrease of ozone at the mid-stratosphere is merely a short term, variation or a start of long term trend, the monitoring of altitudinal profile of ozone is of great significance for the prediction of ozone trend.

REFERENCES

[3.1] Barnola, J. M., Raynaud, D., Korotkevich, Y. S., and Lorius, C., Vostok ice core: a 160,000-year record of atmospheric $CO_2$, Nature, 329, 408-414, 1987.

[3.2] Neftel, A, Moor, E., Oeschfer, H., and Stauffer, B., Evidence from polar ice cores for the increase in atmospheric $CO_2$ in the past two centuries, Nature, 315, 45-47, 1985.

[3.3] Tanaka, M., Makazawa, T., and Aoki, S., Time and space variations of tropospheric carbon dioxide over Japan, Tellus, 39B, 3-12, 1987.

[3.4] Rotty, R. M., and Marland, G., Production of $CO_2$ from fossil fuel burning by fuel type, 1860-1982, Report NDP-006, Carbon Dioxide Information Center, Oak Ridge National Laboratory, 1986.

[3.5] Bolin, B., How much $CO_2$ will remain in the atmosphere? In: The Greenhouse Effect, Climatic Change and Ecosystems, SCOPE 29, B. Bolin et al. eds., 93-155, John Wiley, 1986.

[3.6] Siegenthaker, U., and Oeschfer, H., Biospheric $CO_2$ emissions during the past 200 years reconstructed by deconvolution of ice core data, Tellus, 39B, 140-154,

1987.
[3.7] Houghton, R. A., and Slole, D. L., Changes in the global carbon cycle between 1700 and 1985, In The Earth Transformed by Human Action, B. L. Turner, de., Cambridge University Press, 1990.
[3.8] Tans, P. P., Fung, I. Y., and Takahashi, Y., Observational Constrains on the Global Atmospheric $CO_2$ Budget, Science, 247, 1430-1438, 1990.
[3.9] Raynaud, D., Chappellay, J., Barnola, J. M., Korotkevich, Y. S., and Lorius, C., Climate and $CH_4$ cycle implication of glacial-interglacial $CH_4$ change in the Vostok ice core, Nature, 333, 655-657, 1988.
[3.10] Rasmussen, R. A., and Khalil, M. A. K., Atmospheric methane on the recent and ancient atmosphere; concentration, trend and interhemispheric gradients, J. Geophys. Res. 89, 11599-11605, 1984.
[3.11] Blake, D. R., Rowland, F. S., Continuing worldwide increase in tropospheric methane 1978 to 1987, Science, 239, 1129-1131, 1988.
[3.12] Cicerone, R. J., and Oremland, R., Biogeochemical aspects of atmospheric methane Global Biogeochem. Cycles, 2, 299-327, 1988.
[3.13] Yagi, K., and Minami, K., Effects of mineral fertilizer and organic matter application in the emission of methane from some Japanese paddy fields, Soil Sci. Plant, Nutr., 1990 (in press).
[3.14] Japan Environment Agency, Screening Survey Report on Countermeasures for Global Warming Issue, March 1989 (in Japanese).
[3.15] Prinn, R., Cunnold, D., Rasmussen, R., Simmonds, P., Alyea, F., Crawford, A., Fraser, P., and Rosen, R., Science, 238, 945-949, 1987.
[3.16] Thompson, A. M., and Cicerone, R. J., Possible perturbations to atmospheric CO, $CH_4$, and OH, J. Geophys. Res., 91, 10853-10864, 1986.
[3.17] Khalil, M. A. K., and Rasmussen, R. A., Nitrous oxide; Trends and global mass balance over the last 3000 years, Annal. Glaciol., 10, 73-79, 1988.
[3.18] Rasmussen, R. A., and Khalil, M. A. K., Atmospheric trace gases. Trends and distribution over the last decade, Science, 232, 1623-1624, 1986.
[3.19] WMO/UNEP, Climate Change, The IPCC Scientific Assessment, Cambridge University Press, 1990.
[3.20] UNEP/WMO, Scientific Assessment of Stratospheric Ozone: 1989, Global Ozone Research and Monitoring

Project, Report 20, Geneva, 1989.
[3.21] Tominaga, K., Chlorolfuorocarbons and Stratospheric Ozone , Kagaku, 59, 604-609, 1989 (in Japanese).
[3.22] Makide, Y., Yokohata. A., Kubo, Y., and Tominaga, T., Atmospheric concentrations of halocarbons in Japan in 1979-1986, Bull. Chem. Soc. Japan, 60, 571-574, 1987.
[3.23] Anderson, J. G., Brune, W. H., and Proffitt, M. H., Ozone destruction by chlorine radicals within the antarctic vortex, J. Geophys. Res., 94, 11465-11479, 1989.
[3.24] Japan Environment Agency, Annual Report on the Monitoring of Ozone Layer 1989, March 1990 (in Japanese).

# 4. CARBON DIOXIDE PROBLEMS

## 4.1 COUNTERMEASURES TO THE CARBON DIOXIDE PROBLEM IN HYDROCARBON-FIRED PLANTS

**Yasuo Mori**
Tokyo Institute of Technology
Ohokayama, Meguro-ku, Tokyo Japan

Among the environmental problems discussed in the Introduction, global warming and the restriction of CFC are primarily thermal engineering issues. In particular, global warming, likely to be caused by an increase in the atmospheric carbon dioxide concentration, is one of the most essential and urgent environmental problems. In recent international conferences, held for example by UNEP, a proposal was made that carbon dioxide concentration be controlled under its 1989 level. However, this proposal may not be so forceful, since it is not clear whether the control is to be imposed on each country separately or on the developed countries as a whole. The vague content of the proposal may be attributed to the existing international situation, whereby the energy resources available to each country differ substantially.

In discussions made so far on carbon dioxide problems, atmospheric carbon dioxide is the central issue and a potential greenhouse effect is of great concern to the world people. The relation of the amount of carbon dioxide generated by combustion of hydrocarbon fuel to the increase of atmospheric carbon dioxide suggests that about half the generated carbon dioxide could be absorbed into the seawater. If there is a high possibility for half of it to be absorbed into the seawater, one of the promising measures to control the increase of carbon dioxide concentration is to let it be absorbed into seawater by increasing the rate of diffusion and absorption. Examination of measures using this and other principles for controlling the concentration of atmospheric carbon dioxide leads to the following classification of measures into short, medium, and long-term uses.

(1) Decomposition of carbon dioxide by photosynthesis of chlorophyll.
(2) Separation of carbon dioxide from combustion gas and discharge into the sea.
(3) Combustion of hydrocarbon fuel under a special condition whereby carbon is separated from the fuel during combustion completed by introduction of air.
(4) Separation of carbon from hydrocarbon fuel before combustion and liquefaction of carbon dioxide so that it may be discharged into deep seawater where it will not disturb the ocean environment.

The measure (1) basically depends on the following reaction:

$$n\, CO_2 + n\, H_2O = (H\, CHO)_n + n\, O_2 \qquad (4.1.1)$$

By this reaction, carbon in carbon dioxide is kept in plants as an organic-chemical compound such as $(HCHO)_n$ some of which is resolved again into carbon dioxide and eventually sent out into the atmosphere. In this sense, developing enormous forests would be considered an effective measure against the carbon dioxide problem in the long term. However, from a quantitative analysis on a global scale, the photosynthetic measure is not expected to effectively decrease the atmospheric carbon dioxide concentration as explained in the Introduction, even if the number of plants could be remarkably increased. Nor is the photosynthetic method considered to be a thermal engineering concern. Measure (2) treats combustion exhaust gas with diluted carbon dioxide at atmospheric pressure [1.1.1]. The carbon dioxide separation plant would have to be very large. The low partial pressure of carbon dioxide makes the separation reaction extremely difficult and costly to achieve. The power cost for this method is expected to be about twice as high as full conventional plants. Measure (3) is discussed in detail in the next section. Measure (4) which proposes a new method is explained in detail in this section.

Measure (4) consists of two important processes. One is to separate carbon dioxide from the gas mixture formed from methane fuel and hot steam. The other is to liquefy the carbon dioxide, transport it through a pipe and discharge it into the deep sea through many small holes or jets in the pipe. Except for discharging liquefied carbon dioxide as fine jets into deep seawater, this measure consists mainly of well established technologies. And because the discharge would be at high pressure, into deep seawater where virtually no fish or plants live, this measure would not threaten the ocean environment.

As explained later on, the increase in power costs is calculated to be below 17%. By using measure (4), hydrocarbon fired power plants, steel industries and other factories that consume hydrocarbon fuel and are located on the seashore, can operate without sending carbon dioxide into the atmosphere.

However, motor vehicles and small capacity combustion devices such as those for domestic use, consume hydrocarbon fuel but cannot utilize this measure. In Japan these devices consume about 30% of the hydrocarbon fuel used. Combustors that cannot use measure (4), should control the amount of exhaust gas through effective energy utilization or they should use hydrogen as fuel once safety devices are developed for storage, transportation and combustion of hydrogen fuel.

Fundamental facts giving the basis to measure (4) are explained under the following three items.
(1) The possibility of physically dissolving carbon dioxide in the sea, and chemically converting carbon dioxide to carbonate which would precipitate and settle on the sea bed.
(2) Discharge of liquid carbon dioxide under the sea deep enough to avoid damage to the ocean environment.
(3) The construction, performance and power costs of a plant modified for measure (4) are discussed in the last subsections, 4.1.4 and 4.1.5.

The measure discussed in Section 4.1 cannot be applied to automobiles or domestic equipment. There is no concrete measure for controlling the carbon dioxide emitted in the exhaust gas from automobiles. In my view, the only practical measure is to provide vehicles with engines no larger than the minimum size needed for driving. In other words, just after the oil crisis, many industrialized countries regulated maximum speed and people sought compact cars. Since then many developments in engine performance and devices that precisely control fuel injection rates have been introduced. Owing to these newly developed technologies, a car with a 2 liter engine can run on a highway quietly and safely.

However, because of stable oil costs through the first half of 1990, many people are attracted again to large cars. A car with a 6 liter engine is heavy and sends out about three times more carbon dioxide than a car with a 2 liter engine. The environmental problem related to automobiles should be discussed from the viewpoint stated above. Some degree of regulation on engine volumes may also be required, because about a quarter of hydrocarbon fuel is consumed by cars.

## 4.1.1 Behavior of Carbon Dioxide in the Ocean.

The increased concentration of carbon dioxide in the atmosphere is considered to be caused mainly by combustion in hydrocarbon fueled power plants, steel industries and automobiles and by heating limestone in cement production processes.
From an analysis of the measurement of the increase in atmospheric carbon dioxide at Mauna Loa on the island of Hawaii, U.S.A., about 55% of the carbon dioxide generated in the ways mentioned above is likely to cause the increase of atmospheric carbon dioxide. The rest is supposed to be dissolved in the ocean.
According to geophysical analyses and observations made so far, about 97% of the components of the earth's atmoshpere one hundred million years ago was carbon dioxide. The carbon dioxide present then was gradually absorbed into the ocean and the atmospheric carbon dioxide concentration has slowly decreased down to its present value.
Several reports analyzing this phenomenon say that for about 50 million years the earth has maintained a small and almost constant concentration of atmospheric carbon dioxide. This low concentration level allowed the generation of life on the planet. However, the analyzed results are not necessarily in quantitative agreement with anticipated observations and more accurate models are needed.
The facts explained above are the basis for measure (4)–to discharge liquefied carbon dioxide into the deep ocean. The discharged carbon dioxide would diffuse and mix with the seawater in a short time.
Even though the expectation explained above is considered acceptable, experiments to clarify the phenomena involved would be needed. The global-scale system to be examined includes the absorption of carbon dioxide, the sedimentation of carbonate formed from carbon dioxide and calcium, and the settling of the carbonate on the seabed. More precisely, this system consists of the following processes.
(a) Generation of carbon dioxide in combustion and its diffusion in atmosphere.
(b) Exhalation of carbon dioxide by animals.
(c) Absorption of carbon dioxide by photosynthetic reaction of plants.
(d) Absorption of carbon dioxide from atmosphere into the ocean at the sea surface.
(e) Exhalation of carbon dioxide by marine animals and absorption by marine plants.
(f) Absorption of atmospheric carbon dioxide in rain and its transport to the sea.
(g) Formation of carbonate from ionized carbon dioxide.
(h) Settlement of carbonate on the sea bed.
Of these processes, the absorption of carbon dioxide at the sea surface is physical. Under the surface, the absorption of carbon dioxide by forming carbonate, through chemical reactions of positive ions such as $Na^+$, $K^+$, $Ca^{2+}$, $Mg^{2+}$ with negative carbonate ion $CO_3^{2-}$ and bicarbonate ion $HCO_3^-$ plays a main role.

One of the important factors when discussing the absorption rate of atmospheric carbon dioxide into the deep sea are the physical absorption at the sea surface, which is slow, as well as the absorption of carbon dioxide by the photosynthetic reaction of marine plants and a weak mixing in the vertical direction near the surface: thus, it seems that the small gradient of carbon dioxide concentration near the sea surface controls diffusion of dissolved carbon dioxide into the deeper sea.

The low absorption rate is partially due to the existence of a temperature jump under the tidal zone where a convection flow is hardly expected other than molecular diffusion. One piece of evidence that supports the presumption above, is a report indicating that it takes approximately forty years for the carbon dioxide physically absorbed at the sea surface to naturally settle at a zone about 200 m deep. The layer between the surface and the 200 m depth zone is called the tidal layer where the tidal flow plays a dominating role, and causes low mass transfer in the vertical direction. The long period of several hundred million years that it took for the dense carbon dioxide in the atmosphere to be absorbed in the seawater and reduced to its present concentration level, is considered to be attributed to the slow rate of mass transfer. Therefore, it is considered that because of this low mass transfer rate, if carbon dioxide is discharged in a liquid or solid phase into the sea less than 200 m in depth, carbon dioxide hardly goes down to become carbonate and settle on the seabed.

As explained above, the carbon dioxide should be discharged under the sea at a zone both deep and cool enough to form carbonate which precipitates and settles on the seabed without causing any damage to the aquatic environment. This is discussed in Subsection 4.1.2.

Indicating the density of carbonate ion as $\{CO_3^{2-}\}$ (kmol/m$^3$), that of bicarbonate ion as $\{HCO_3^-\}$, that of physically dissolved carbon dioxide as $\{CO_2\}$, and the total density of dissolved carbon dioxide as $\{\Sigma CO_2\}$, a measured result of these components [4.1.1.1] is shown on Fig. 4.1.1.1 On this figure, the distribution of the pH of seawater is also shown. $\{H^+\}$ can be calculated by the following equation and pH is expressed in Equation (4.1.1.3) by use of $\{A\}$ defined below.

$$HCO_3^- \rightarrow CO_3^{2-} + H^+ \qquad (4.1.1.1)$$

$$\{A\} = \{HCO_3^-\} + 2\{CO_3^{2-}\} \qquad (4.1.1.2)$$

$$pH = -\log\{K_2 \frac{2\{\Sigma CO_2\} - \{A\}}{\{A\} - \{\Sigma CO_2\}}\} \qquad (4.1.1.3)$$

$K_2$ in Equation (4.1.1.3) is the equilibrium constant and was obtained by experiments. Measurements of carbon dioxide off of Akuni Island in the Okinawa area of Japan [4.1.1.1] indicate that the difference between the concentrations of carbon dioxide at the 980 m depth and the surface is 0.25 mol/m$^3$, while the value calculated from the oxygen decrease is 0.28 mol/m$^3$. This could reflect photosynthetic activity in the upper layer of the ocean. Moreover, it is estimated that two thirds of the calcium carbonate generated settles on the sea bed. It may be noted here that limestone and marble are sediments made of carbonate.

The preceding discussion could indicate the advantage of disposing of liquid carbon dioxide into the deeper sea. A concrete method to implement this scheme is explained in Subsection 4.1.3. As seen on Fig. 4.1.2.1, the sea can

As indicated on Fig. 4.1.2.1, turbulent mixing in the shallow zone (less than 100 m), makes the temperature and nutrient distributions uniform because of a forced convective flow such as caused by tidal currents. Therefore, when carbon dioxide is discharged in the shallow zone (less than 100 m), it is likely to evaporate and mix with the atmosphere.after reaching the sea surface. The 100 m depth is too shallow for discharging liquefied carbon dioxide. An analysis of nutrient distribution shows that in the forced convective zone, plankton, on which many species of fish feed, thrive and consume nutrients at a high rate. On the other hand, in the zone below 500 m, due to a weak free convective flow, little plankton or fish are found, and the distribution of nutrients are nearly uniform as well. The zone below 500 m is quite different from the shallow zone from the fluid mechanical and thermodynamical viewpoints. The temperature for the maximum density of seawater decreases with an increase in salinity. Fig. 4.1.2.2 shows the relationships between the temperature for maximum density and the freezing point versus salinity. The maximum density occurs at freezing when salinity reaches 2.5%. Cold water mainly from the the South Pole region flows along the seabed. This cold water mixes with ambient seawater while flowing toward the Equator, and the temperature and salinity gradients turn it into a large scale, free convective flow. As seen on Fig. 4.1.2.1, the zone below 500 m where the distributions of temperature and nutrients are almost uniform is thus called the free convective zone in this section. Carbon dioxide in the deeper zone is not likely to ascend through the relatively undisturbed temperature jump layer to the shallow zone and mix with the atmosphere. In other words, in order to prevent liquid carbon dioxide from reaching the sea surface and the atmosphere, it should be discharged below 500 m in the free convective zone. An implementation of this measure is explained in Subsection 4.1.3.

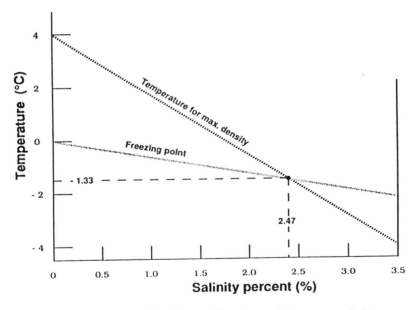

**FIGURE 4.1.2.2** Maximum density and freezing point versus salinity.

Before proceeding, however, marine environment should be considered. Most fish move in specific tidal currents and dwell in the zone shallower than 200 m which contains both a variety of planktonic organisms and a great deal of oxygen dissolved from the atmosphere. In this zone, carbon dioxide is also dissolved from the atmosphere and solar intensity is strong enough for marine plant photosynthesis (plants absorb carbon dioxide and exhale oxygen).

However, solar intensity exponentially decreases with depth and below 200 m there is very little sunlight. Consequently, almost all activity related to marine life occurs above 200 m. Below about 500 m, there is little life. The forced convective and free convective zones are separated by the so-called "temperature jump layer" and the discharge of liquid carbon dioxide in the deeper, free convective zone is not likely to damage the marine environment. As explained above, out of the conditions listed at the beginning of this subsection, (1) and (2) are already satisfied by the depth requirement specified above. Conditions (3) and (4) will be discussed in the next subsection; (5) and (6) will be addressed in Subsection 4.1.4.

### 4.1.3 Discussion of the Discharge of Liquid Carbon Dioxide into the Deep Sea from a Thermodynamic Standpoint.

On Fig. 4.1.3.1, a thermodynamic diagram for carbon dioxide is shown. For pressures below 5.18kgf/cm$^2$ ($\approx$ 500 kPa), only vapor and solid phases exist, while above that threshhold, liquid and vapor can be observed. The discharge of large quantities of solid carbon dioxide into the sea is not advisable, as it evaporates and the solid would go up to the sea surface. The discharge of large amounts of solid carbon dioxide divided into small quantities is neither simple nor economically feasible. In addition, the solidification of carbon dioxide needs a great deal of power. In hydrocarbon-fired power plants or other thermal energy industries, methane and coal-gasified fuel are thought to be the important energy sources in the near future. Coal-gasified fuel may be one of the most promising alternatives with the rising cost of oil and the use of nuclear energy developing slowly. The main components of coal or hydrocarbon gasified fuel are methane, hydrogen and carbon monoxide, with the exception of coal gasified fuel of low combustion heat. Methane is becoming widely used in power plants because it produces less carbon dioxide. In the following discussion, methane will be examined as a fuel source.

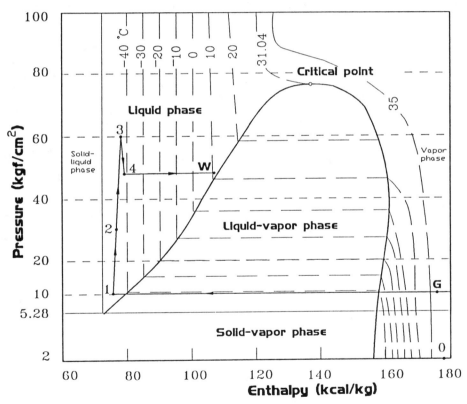

FIGURE 4.1.3.1 Thermodynamic Diagram for Carbon Dioxide.

When considering piping liquid carbon dioxide into the deep sea, the steps necessary for processing hydrocarbon fuel to prevent any carbon dioxide from escaping into the atmosphere are:
(1) Before burning hydrocarbon fuel such as methane, a reforming reaction between reheated methane and hot steam on a catalyst is performed to produce hydrogen and carbon dioxide.
(2) The separation of carbon dioxide from the reformed mixture gas is made in a chemical reactor by means of a carbon dioxide absorbing liquid at a pressure of about 2 to 20 atmospheres as described below. Hydrogen, thus obtained, is used as fuel for the plant. Hydrogen flows via a short pipe so that safety problems are avoided.
(3) Separated carbon dioxide is then liquefied and transported through a long pipe into the deep sea.

The methane-steam reforming technology, which constitutes the technological basis for this new proposal, is a well established and economically feasible method of producing hydrogen. In addition, the technology to separate carbon dioxide and hydrogen from fuel is also widely employed, using appropriate pressures in a carbon dioxide absorbing tower and a regenerating tower, respectively. The optimum pressure in the absorbing tower is about 20 atmospheres, which is consistent with that required for the reforming reaction, while the pressure in the regenerating tower is about 2 atmospheres. Those are the required pressures for operating the separation process economically. A currently used method in a hydrocarbon-fired plant for separating carbon dioxide from the exhaust gas at 1 atmosphere is not economically feasible because of two serious disadvantages: the difficulty of treating large flow rates of flue gas, and the use of inadequate pressures resulting in a large power consumption. On Fig. 4.1.3.1, the thermodynamic state of carbon dioxide is shown, which, after leaving the regenerating tower, is cooled by exchanging heat with incoming air. In order to liquefy the carbon dioxide, available at 2 atmospheres and about 40°C, it is compressed to 10 atmospheres and cooled. The state thus reached is shown as G on the figure. Carbon dioxide in state G is cooled in a refrigerator down to point 1 at about -45°C and liquefied carbon dioxide is obtained. Liquefied carbon dioxide thus treated is transported

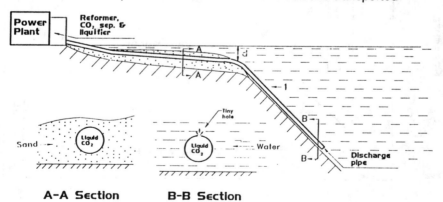

FIGURE 4.1.3.2  Model diagram for transporting liquefied carbon dioxide into the deeper sea.

through a pipe down into the deep sea as seen on Fig. 4.1.3.2. The state shown by a letter or a number on Fig. 4.1.3.2 is shown on Fig. 4.1.3.1 by the same symbol. The discharge process is explained by use of Figs. 4.1.3.1 and 4.1.3.2. Point 1 on Fig. 4.1.3.2 indicates the exit from the refrigerator. When a fossil-fired power plant is, for example, located a short distance away from the shoreline, the pipe is protected by a thermal insulating layer, and from the seashore down to about 200 m depth, it is buried several meters in the seafloor. The diameter of the pipe depends on the power output of the plant and is in the order of a few tenths of one meter for a several hundred MW plant. The technology to bury such a pipe in sand has been well developed and resorts to marine robotics. The thermal conductivity of sand depends much on its dampness, kind and grain size. Burying the pipe eliminates interference with dragnet fishery, and prevents corrosion. The heat input to the pipe from the sea surface down to the 200 m depth indicated by point 2 barely heats up liquid carbon dioxide by a few degrees. From point 2 down to point 3 at the 500 m depth, the carbon dioxide temperature increase depends on the length of pipe, that is on seabed slope; it is estimated within a few degrees, because the seawater outside the pipe is almost still and the heat transfer coefficient on the outer pipe surface is rather low. Therefore, the overall temperature increase for liquid carbon dioxide is at most ten degrees and the thermodynamic state at the pipe exit is shown by 4. Even when the liquid carbon dioxide is heated up to the water temperature at 500 m depth, it is always in liquid phase as understood from Fig. 4.1.3.1. The specific gravity of liquid carbon dioxide is about 1.07 at -40°C and the pressure difference between carbon dioxide and outside seawater at a given depth is always about 10 atmospheres. In the proposal made in this Section, no discharge of liquid carbon dioxide is performed above 500 m; below that depth, the pipe has many tiny holes about 10 mm in diameter bored with an axial pitch of a few meters and distributed in a spiral way along the axial direction [4.1.3.1]. Discharge is sustained by the pressure difference of about 10 atmospheres. The jet of liquid carbon dioxide breaks down into many small droplets which are supposed to be initially surrounded by an ice layer, which should melt quickly. The boiling of liquid carbon dioxide never occurs. Thus, carbon dioxide diffuses mainly through turbulent and molecular diffusion induced by the slow convective flow of seawater. On Fig. 4.1.3.1, point 3 indicates the state just before discharge and point 4, the state after discharge. Carbon dioxide at point 4 mixes with water in state W. The discharge and diffusion are performed at pressures of about 50 atmospheres. Since no experiments have been reported, some studies on discharge of liquid carbon dioxide jets are desirable. If a small number of fish or plant species live in such a deep area, they will not be disturbed by water containing extremely diluted carbon dioxide. The measure proposed and explained in this section has a promising technical feasibility and is provided with the following features.

(1) Liquefied carbon dioxide is discharged through many tiny holes bored in the pipe so as not to generate areas of high carbon dioxide concentration in the sea. Liquid carbon dioxide does not reach the sea surface. In contrast to the natural process under which carbon dioxide dissolved solved into the sea from the atmosphere takes about 50 years to go down through the forced convective zone, it is here artificially transported in a short time directly to the free convective layer below 500 m.

(2) Carbon dioxide is discharged little by little through many tiny holes bored in the pipe into the sea below the 500 m depth. Therefore, almost no damage

is done to marine life, because most species live in waters shallower than 200 m.
(3) Carbon dioxide is obtained by a reforming reaction between hydrocarbon fuel and hot steam before combustion, and hydrogen thus obtained is used as fuel; consequently, no carbon dioxide is contained in the exhaust gas of the plant.
(4) The pipe for transporting liquefied carbon dioxide could be buried in the seabed if interference with dragnet fishery poses a problem.

Reference:
[4.1.3.1] Mori, Y., Combustion Method Containing Least Carbon Dioxide in Exhaust Combustion Gas, Japanese Patent Application 01-0673585, 1988.

## 4.1.4 Separation of Carbon from Hydrocarbon Fuel before Combustion in a Steam Power Plant.

The environmental problem caused by the increase of carbon dioxide concentration in atmosphere has recently been considered on a global scale. Up to now, most of the thermal energy consumed by human beings has been generated by hydrocarbon fuel combustion. One of the countermeasures to the problem reported so far is to separate carbon dioxide from the exhaust gas of a fossil-fired plant and discharge it into the sea [4.1.4.1]. The rough estimate of power cost, including the costs of carbon dioxide separation equipment and discharge pipe, is reported to be as much as twice the conventional power cost. Among the various combustors of hydrocarbon fuel, those of small size or of a mobile type, as on the automobile, are not suitable for the application discussed in this section. Most of the fossil-fired steam power plants and steel industries in Japan are located close to the seashore for easy transportation of fuel and products. Thus, the measure discussed in this subsection can be effectively used in these large and stationary plants. When a country has an extensive coastline, the government may press a request to build hydrocarbon-fueled plants on the shore. The measure proposed in this section is considered to make a significant contribution to the control of the increase in carbon dioxide concentration on a worldwide scale. Recent international discussions on the regulation of the exhaust amount of carbon dioxide suggest that it is becoming a serious problem.

Every industrialized country will have to seek countermeasures or methodologies to decrease the amount of carbon dioxide exhaust by about 10 to 20% within ten years. The present measure is proposed with due consideration of the disadvantage that separation of carbon dioxide from the combustion flue gas would require the processing of large volumes because the concentration of carbon dioxide is low. When coal or liquid fossil fuel is used, it is gasified to extract sulfur. The gas fuel thus obtained is mainly methane and the rest is mostly hydrogen. Therefore, there is little loss of generality in discussing methane as the only fuel used. In the method proposed, a gaseous mixture of methane and steam preheated up to about 700°C is reformed on a catalyst to produce carbon dioxide and hydrogen. The carbon dioxide is separated from the mixture in carbon dioxide absorbing and regenerating towers by use of a carbon dioxide absorbing liquid. Thus, hydrogen instead of hydrocarbon fuel is burned in the boiler to generate high temperature superheated steam which drives the turbine. The exhaust gas from the plant consists of nitrogen and steam produced by the combustion of hydrogen and air. Separated carbon dioxide is led to the liquefaction equipment after being cooled down by incoming air. Liquefied carbon dioxide is transported through a long pipe down into the sea and discharged there to prevent any damage to the marine environment. The plant consists of the following four parts:
(1) Steam generator (boiler).
(2) Reformer of hot methane and steam.
(3) Carbon dioxide absorbing and regenerating equipment.
(4) Liquefier of carbon dioxide.

In comparison with the method reported previously about the separation of carbon dioxide from combustion exhaust gas, the newly proposed measure

described in this chapter has the following features:
a. The reformer to produce the mixture of carbon and hydrogen from hot methane and steam can be installed in the boiler duct at a location appropriate for the reforming reaction and no additional burner or duct is necessary.
b. The methane preheater and steam generator for reforming are operated by combustion gas and reformed mixture gas.
c. Only hydrogen is burned in the boiler combustor and the concentration of NOx is thinner than when hydrocarbon fuel is burned, which represents an additional environmental advantage.
d. Methane is treated before combustion, therefore carbon dioxide separation equipment can be small.

The reforming pressure is almost the same as that of the carbon dioxide separation process and the power consumed for carbon dioxide separation is rather low. One of the model systems illustrating the measure is shown on Fig. 4.1.4.1. The performance and construction of a conventional boiler for a hydrocarbon-fired power plant is well known, so that only the reformer and the relation of the reformer to the boiler are explained below by use of Fig. 4.1.4.1. At the bottom of the boiler furnace, the combustor is installed to burn hydrogen and preheated air. The reformer made of thermal-resistive superalloy contains a nickel catalyst to allow a reforming reaction between pre-heated methane and steam at about 800°C. It is installed at a position in the boiler furnace where the combustion gas temperature is about 1100°C. A methane preheater to heat methane up to about 500°C is installed outside the boiler duct to use the hot mixture gas from the reformer, which also releases its heat to the preheated steam and cools down to about 200°C before being sent to the carbon dioxide absorbing and regenerating towers. The mixture gas is cooled further by the carbon dioxide rich liquid to about 130°C. In the carbon dioxide absorbing tower only carbon dioxide is absorbed into the liquid and hydrogen is thus separated. The liquid rich in carbon dioxide is sent to the regenerating tower after being depressurized to about 2 atmospheres and heated by the mixture gas from the methane-and-steam preheater. Carbon dioxide is released as a gas, which, upon leaving the regenerating tower, is cooled by air incoming to the plant for combustion. The heat balance for the system is discussed below. In the reforming reaction, the proper stoichiometric ratio of steam to methane is 2 as shown in Eq. (4.1.4.1), but, in general, it may be larger, typically about 4.

$$CH_4 + 2H_2O = 4H_2 + CO_2 \qquad (4.1.4.1)$$

This chemical reaction is endothermic, and consumes 165.1 MJ/kmol($CH_4$). The combustion heat of $4H_2$ is 1132.7 MJ/kmol($CH_4$). Therefore, the heat available for producing superheated steam as in a conventional boiler is 967.6 MJ/kmol($CH_4$). The heat value of four hydrogen molecules is larger than that of one methane molecule and the difference is used to generate the heat necessary to sustain reforming reaction. One important problem to be addressed is the heat required to bring reforming water to the saturated steam condition. This heat is provided by lowering the temperature of the exhaust gas and by increasing the flow rate of hydrogen. If only the latter were possible, the thermal efficiency of the power plant provided with the equipment treating carbon dioxide would decrease by about 2 %. The optimum reforming pressure is known to be about 20 atm, and methane is compressed to this pressure in advance. The pressure in the reformer down to the carbon dioxide absorbing

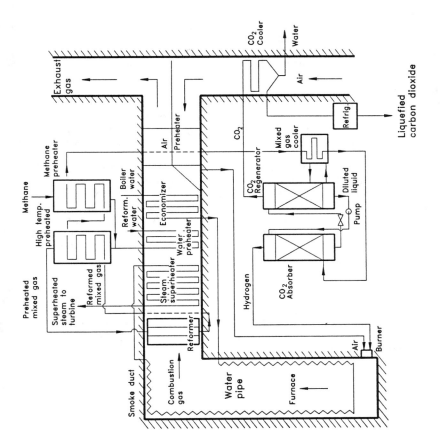

FIGURE 4.1.4.1  Model flow system of boiler and reforming part.

tower is about 20 atmospheres, and the optimum pressure in the regenerating tower alone is about 2 atmospheres. It is again noted that the exhaust gas of this power plant contains less nitrogen oxide due to the fact that little NOx is generated in the combustion of hydrogen [4.1.4.1], [4.1.4.2]. The features of the measure proposed in this section are summed up as follows.

(1) In advance of combustion, preheated methane and steam are reformed on a catalyst to produce hydrogen and carbon dioxide. Hydrogen is separated from carbon dioxide and is used as a fuel for the boiler. Carbon dioxide thus separated is liquefied, transported into the sea below the 500 m depth through a pipe and discarded from many tiny jets to diffuse, dissolve and form sediments, without causing environmental damage or increasing atmospheric carbon dioxide.

(2) In contrast to the proposal reported earlier to separate carbon dioxide from the combustion exhaust gas, the gas flow rate required here is one order of magnitude smaller, and the installation of a reformer for producing carbon dioxide and hydrogen from methane and steam in the boiler of a steam power plant or a furnace is readily possible. Therefore the cost of the new equipment is low.

(3) Carbon dioxide dissolved in the deep sea reacts with calcium ions to become carbonate and finally settles on the seabed.

(4) The reformer is not required to produce hydrogen of high purity and may be operated with fuels of different properties and under various conditions.

(5) The amount of hydrocarbon fuel on this planet is expected to be exploitable over several hundred years. The measure proposed in this chapter may be understood as a mean to open the way to utilize hydrocarbon fuel for the future without causing any damage to the global environment.

(6) For power plants and industries which are located by the seashore and burn much fossil fuel, the measure proposed in this subsection is easily applicable to drastically reduce the carbon dioxide concentration of the exhaust combustion gases.

References:
[4.1.4.1]  Y. Mori, Combustion Method and Combustor with Little Carbon Dioxide Exhaust , Japanese Patent Application, Hei 1- 023664.
[4.1.4.2]  T.Miyauchi,Y.Mori and T.Yamaguchi, Effect of Steam Addition on $NO_x$ Formation, The Combustion Institute (International Symposium on Combustion), p. 41,1981.

| | | | | | | | | |
|---|---|---|---|---|---|---|---|---|
| 20 | $C_4H_4+C_2H$ | $=C_4H_3+C_2H_2$ | Q6 | | | 3.98D13 | 0.0 | 0.0 | [17] |
| 21 | $C_4H_3+M$ | $=C_4H_2+H$ +M | Q7 | | | 1.00D16 | 0.0 | 60.0 | [14] |
| 22 | $C_4H_2+M$ | $=C_4H+H$ +M | Q8 | | | 3.46D17 | 0.0 | 80.03 | [17] |
| 23 | $C_2H+C_2H$ | $=C_4H+H$ | Q65 | | | 1.81D13 | 0.0 | 0.0 | [4] |
| 24 | $C_2H_3+C_2H$ | $=C_4H_4$ | Q66 | | | 1.81D13 | 0.0 | 0.0 | [4] |
| 25 | $C_2H_3+C_2H$ | $=C_4H_3+H$ | Q67 | | | 1.81D13 | 0.0 | 0.0 | [4] |
| 26 | $C_4H_2+H$ | $=C_4H+H_2$ | Q12 | | | 1.00D14 | 0.0 | 20.0 | [20] |
| 27 | $C_4H_4+M$ | $=C_2H_2+C_2H_2+M$ | Q39 | 2.28 | 2.35 | 5.60D19 | 0.0 | 84.5 | [21]·¹ |
| 28 | $C_2H+C_4H_2$ | $=C_2H_2+C_4H$ | Q40 | | | 2.00D13 | 0.0 | 0.0 | [16] |
| 29 | $C_2H+C_4H_4$ | $=C_4H_2+C_2H_3$ | Q41- | | | 1.00D13 | 0.0 | 0.0 | [16] |
| 30 | $C_2H_3+C_4H_4$ | $=C_2H_4+C_4H_3$ | Q43 | | | 5.01D11 | 0.0 | 16.3 | [22] |
| 31 | $C_2H+C_2H_4$ | $=C_4H_4+H$ | Q44 | | | 3.01D11 | 0.0 | 3.82 | [16] |
| 32 | $C_4H_4+M$ | $=C_4H_2+H_2+M$ | Q45 | 0.61 | 2.74 | 2.40D19 | 0.0 | 84.5 | [21]·² |
| 33 | $C_4H_3+H$ | $=C_4H_2+H_2$ | Q46 | 1.01 | 1.00 | 1.00D13 | 0.0 | 0.0 | [22] |
| 34 | $C_2H+C_2H$ | $=C_4H_2$ | Q21- | | | 1.81D13 | 0.0 | 0.0 | [4] |
| 35 | $C_4H_4+H$ | $=C_4H_3+H_2$ | Q47 | 0.97 | 1.06 | 1.50D14 | 0.0 | 10.2 | [19] |

·¹ Q39 in case E' 4.26D18 0.0 74.8
·² Q45 in case E' 2.23D18 0.0 76.0

# A,n,E are the parameters of rate expression $k=A \cdot T^n \cdot \exp(-E/RT)$ for the forward direction. The reverse rate constants are calculated via equilibrium constants given in Table 4.2.1.

### 4.2.2 Result

The computed results are compared with the experimental values of $C_4H_4$, and $C_4H_2$ obtained by shock-heated sample gases of 5 % in argon at total density $\rho_5=2.35E-5$ mole/cm³ and reaction time of 1 msec. To realize the reasonable coincidence with the experimental yield of $C_4H_4$ at 1500K, some rate constants of sensitive reactions in Table 4.2.2 are modified by multiplying adjusting factors. The set of sensitive reactions was introduced by the sensitivity analysis carried out beforehand. Sensitivity coefficients of most of the reactions are quite small and indicate the existence of equilibration. As an example some of the coefficients calculated in the case E are shown in Table 4.2.2, where the coefficients within $1\pm0.01$ are omitted. The sensitivity coefficient of reaction i is defined here simply by the ratio of the calculated yields $Y_5$ and $Y_0$ as,

$$S_i = Y_5/Y_0$$

where $Y_5$ and $Y_0$ are the $C_4H_4$ (or $C_4H_2$) yields computed for the reference conditions but with the rate constants for reactions i (including reverse reaction) multiplied by a factor of 5 and those with the unmodified rate parameters, respectively.

However, it was unable to obtain a consistent result in the case of A unless modifying adjusting factors more than two orders of magnitude. This means the value of 114 kcal/mole for the heat of formation of $C_2H$ is unsuitable. On the contrary, in the cases of B to E, it is possible to simulate the maximum $C_4H_4$ yield at 1500K by the slight change of the adjusting factors in each case.

Table 4.2.3 shows the values of the adjusting factors multiplyed to the rate parameters of Table 4.2.2 to obtain the nearly coincidence with the observed yields at 1500K.

In Fig.4.2.1 the computed results on the case of E was plotted as an example, where ordinate represents the logarithm of yields in ppm unit.

TABLE 4.2.3  Adjusting factors to match the observed yields

| code | case B | case E | case E' | case F |
|---|---|---|---|---|
| D25 | 1.0 | 1.0 | 1.0 | 0.5 |
| Q1  | 1.0 | 1.0 | 1.0 | 0.5 |
| Q39 | 1.0 | 0.8 | 0.55 | 1.0 |
| Q45 | 0.8 | 1.2 | 0.8 | 1.0 |
| Q82 |     |     |     | 0.5 |

As clearly shown in Fig.4.2.1, the calculated profile shifts to the lower value at the lower temperature side. According to the recent paper written by Kiefer,von Drasek[4.2.23] on the same subject, an excellent coincidence was obtained in the whole temperature range. To reproduce this agreement we construct a simplified scheme F, in which chemical species other than those of Table 4.2.1 are excluded, since those species such as $C_4H_6$ and higher hydrocarbons $C_6-$, $C_8-$ are considered merely as sinks of highly unsaturated hydrocarbons. So that reactions R20~25, and R37~40 of their scheme are excluded, and only the one-way reactions R15, R19, R30 and R34 (the rate parameters of the reverse reaction of R34 are obtained from Colket[4.2.24]) are taken into account. The remaining 26 reactions are considered to be reversible and the reverse rate constants are determined by detailed balance. The forward rate constants used in the scheme F are the same as those given by Kiefer,von Drasek except R2, R3 and R13. The parameters of the latter three are reduced to 1/2 of Kiefer's value (see Table 4.2.3) to obtain the similar satisfactory results as represented in Fig.4.2.1. Being stimulated by this success we changed the activation energies of Q39 and Q45 in our scheme E to the lower values of 74.8 and 76kcal/mole, respectively. The modified scheme is named as E' and the computed results were considerably improved.

The most characteristic difference between scheme E' and Kiefer's scheme (including scheme F) is concerned with the participation of Q82 (Kiefer's R2). In the latter scheme Q82 is treated as one of a core molecular process, but since we have no information about the reverse step, we omit Q82 [$C_2H_2+C_2H_2=C_4H_2+H_2$] from our scheme B~E'. On the contrary, the other molecular process Q39 [Kiefer's R1: $C_4H_4=C_2H_2+C_2H_2$] was included in scheme B~E', since the reversibility of this process was confirmed by Kiefer,Mitchell[4.2.25].

Conclusion
As the heat of formation of $C_2H$ the higher value of 132~135 kcal/mole is preferable to the lower value of 114 kcal/mole, but it is impossible from the present simulation to settle the choice between 154 and 186 kcal/mole for the heat of formation of $C_4H$.

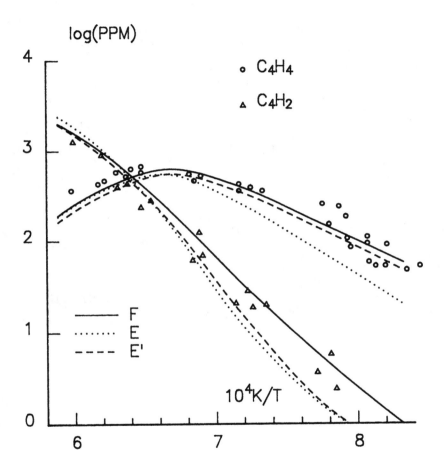

FIGURE 4.2.1 Product distribution as a function of temperature

Reference
[4.2.1] M.Frenklach,D.E.Bornside:Combustion Flame,56(1984)1.
S.J.Harris,A.M.Weiner +1:Combustion Flame,72(1988)91.
[4.2.2] H.Ogura:Bull.Chem.Soc.Japan,50(1977)1044.
[4.2.3] J.Warnatz:"Combustion Chemistry" Chap.5,ed. by W.C.Gardiner (1984),Springer.
[4.2.4] W.Tsang,R.F.Hampson:J.Phys.Chem.Ref.Data,15(1986)1087,16(1987)471.
[4.2.5] P.R.Westmoreland,B.Howard:21st Intn.Combustion Symposium,773 1986).
[4.2.6] JANAF Thermochemical Tables 2nd ed.(1971),3rd ed.(1984).
[4.2.7] R.Atkinson et al:J.Phys.Chem.Ref.Data,18(1989)881.
[4.2.8] E.Ghibaudy,A.J.Collussi:J.Phys.Chem.,92(1988)5839.
[4.2.9] K.Kuratani:Rep.Inst.Space Astron.Sci.,No.58(1986).
[4.2.10] K.Kuratani:Res.Note Inst.Space Astron.Sci.No.320(1986).The revised edition is compiled in a floppy-disk form(1990).
[4.2.11] T.Tsuboi:Japan J.Appl.Phys.,17(1979)709.
[4.2.12] J.H.Kiefer,S.A.Kapsalis,Combustion Flame,51(1983)29.
[4.2.13] S.W.Benson,G.R.Haugen,J.Phys.Chem.,71(1967)1735.
[4.2.14] C.K.Westbrook,F.L.Dryer +1:19th Intn.Combustion Symposium,153 (1983).
[4.2.15] M.Frenklach,S.Taki +2:Combustion Flame,54(1983)81.
[4.2.16] M.Frenklach,D.W.Clary:20th Intn.Combustion Symposium,887( 1984).
[4.2.17] T.Tanzawa,W.C.Gardiner:J.Phys.Chem.,84(1980)236.
[4.2.18] D.B.Olson,W.C.Gardiner:Combustion Flame,32(1978)151.
[4.2.19] M.Frenklach,J.Warnatz:Combustion Sci.Tech.,51(1987)265.
[4.2.20] J.Warnatz:Combustion Sci.Tech.,34(1983)177.
[4.2.21] Y.Hidaka,T.Nakamura +3:Intn.J.Chem.Kinetics,21(1989)643.
[4.2.22] M.B.Colket:21st Combustion Symposium,851(1986).
[4.2.23] J.H.Kiefer,W.von Drasek:Intn.J.Chem.Kinetics,22(1990)747.
[4.2.24] M.B.Colket,D.J.Seery:Combustion Flame,75(1989)343.
[4.2.25] J.H.Kiefer,K.I.Mitchell +2:J.Phys.Chem.,92(1988)677.

# 4.3 A NEW APPROACH TO REDUCE $CO_2$ EMISSION BY RADIATION-CONTROLLED COMBUSTION AND CARBON RECOVERY

**Ryozo Echigo**
Department of Mechanical Engineering
Tokyo Institute of Technology
Ohokayama, Meguro-ku, Tokyo 152, Japan

## Introduction

In view of the increasing evidence based on the scientific measurements and prediction, the climate changes on global scale have been discussed extensively, in particular, related to the carbon dioxide emission into atmosphere through the consumption of the fossil fuel [4.3.1]. It is also pointed out that the anthropogenic emissions of gases other than $CO_2$ (so called greenhouse gases; CFCs, $O_3$, $N_2O$, $CH_4$, $H_2O$ • • •) are likely to contribute appreciably to the global warming.

The concentration of the carbon dioxide in the atmosphere is controlled by the so many natural phenomena such as the absorption and/or emission of $CO_2$ into and/or from the ocean water, the adsorption into or release from the terrestrial sphere, the photosyntheses or the deforestation, and so forth. It is a grave concern that the amount of the carbon dioxide emission is approximately 22 billion ton/year, while the annual increase of the carbon dioxide in the atmosphere is estimated to be 10 billion tons (about 2 to 3 % of the total amount of the carbon recirculation into and/or out from the atmosphere).

We have to be aware of the fact that the atmosphere has a certain capacious limit for the emission of the greenhouse gases. On the other hand, leaving room for further stimulative discussions, the fossil fuels are generally recognized to be a major feedstock for a subsequent few decades in the light of the economical and technological feasibility. In this respect a new energetics would be awaited, on a high priority basis, concerning the entire system of the combustion technologies so as to realize the substantial mitigation of the global environmental threat.

In order to make a breakthrough in such difficulties, it is proposed a new concept, i.e. tentatively defined here to be "Carbon Solidification Combustion" (designated as **CSC** hereafter); the concept is not amenable to the traditional one on which, for many years, the heating values of the fossil fuels are based. Since the **CSC** is so defined as to control the combustion to recover the carbon in solid phase and to utilize the residual energy, the virtual energy release during the **CSC** processes decreases compared with an original fuel. The validity of the **CSC** will be endorsed through a brief examination on the dry ice fabrication process in which the carbon dioxide in flue gases is converted to the solid (dry ice) in order to abandon into the deep sea. The subsequent discussion will be extended to elucidate a possible criterion for the **CSC** by the basic experiment on the soots producing combustion.

## A new combustion technology - carbon solidification combustion (CSC) -

Over a past decade, by the experts in the workshops of WMO, UNEP, IPCC, etc., there have been discussed and proposed a number of strategies for reducing the greenhouse gases, particularly $CO_2$; among of them some typical approaches are classified from the

technological viewpoints and are illustrated in Fig.4.3.1, together with the **CSC**. Subjecting to the simple examination they may possibly fall into five broad categories;
(1) promotion of the energy saving and improvement of the efficiency of the energy equipments and facilities,
(2) shift from the fossil energy to the non-fossil energy (nuclear and natural energy),
(3) conversion from the high C/H ratio fuel to the low C/H ratio fuel (i.e. natural gas),
(4) store and/or abandon to the deep sea or the exhausted mines after liquidized or solidified,
(5) enhancement of the forestation.

Among these categories, only (3) and (4) allow the use of the fossil fuels. Concerning category (3), it should be noted that there are many proposals in which $CO_2$ is converted to some other resources (methane, alcohols, etc.) without specifying the energy sources so that the effectiveness is somewhat dubious. It is also worth noticing that in many cases the amount of the recovered resources turns out to be just bulky compared with the consumption (supply-demand mismatch).

Provided that the fossil fuel is a primary energy, there are three options as indicated in the following:
(a) reform of the fuel *prior to* the combustion
(b) control of the $CO_2$ emission *during* the combustion processes
(c) store or abandon the $CO_2$ through liquidized or solidified from the exhaust gases *after* the combustion

Besides, the energy needed in the preceding physical and chemical processes has to be self-sustaining, otherwise it is not regarded to be a primary energy. In terms of these options, the categories (3) and (4) are regarded as (a) and (c), respectively. Unlike these categories, the **CSC** presented here belongs to the option (b).

An outline of the **CSC** is that some part of the carbon in fossil fuel is fixed (solidified) as the soots or the cokes during the combustion, and the residual carbon is burned together with the hydrogen. In the **CSC** it is noted that while the emission of $CO_2$ could be reduced with increasing carbon solidification ratio, the substantial heating value of fuel is also decreased.

A simple speculation may criticize that the **CSC** is too radical on the basis of the conventional concept that an essential part of fossil fuel consists of the carbon. It is, however, worthy to remind of the fact that if all the carbon is burned (the category (4)), the additional energy is necessary in order to reduce the $CO_2$ emission into the atmosphere. Furthermore, the conversion to low carbon fuels (e.g. natural gas) indicated in the category (3) is seemingly quite reasonable but this idea does not account for the utilization of the most important fossil resources such as coal and petroleum. In this meaning the context of the category (3) is much more radical rather than the **CSC**, because it implies the heavier carbonaceous fuels to be not in use.

In this regard it is repeatedly emphasized that entirely new concepts have to be introduced in the broad lines of disciplines (not only in the combustion science and engineering but in some other fields). The feasibility of the **CSC** is, consequently, wide open for intensive discussion in order to obtain a mutual consent that a drastically difficult strategy is necessary for overcoming the catastrophic circumstances never experienced before.

As being discussed later in more details, the **CSC** will be compared with the countermeasure (4) in which the comprehensive discussions have been done by evaluating the energy cost of the electric power generation, and by taking account of the energy required for separation, pressurization and refrigeration [4.3.2, 3]. Finally, it is interesting to note that M.Steinberg, who proposed an idea of dry ice fabrication from $CO_2$ for submerging into the deep sea [4.3.2], has presented an opinion similar to the **CSC**. In his recent paper [4.3.4] it is proposed that the coal is cracked into its elements of C and $H_2$, and then the former (carbon black) is returned to the mine for possible future use.

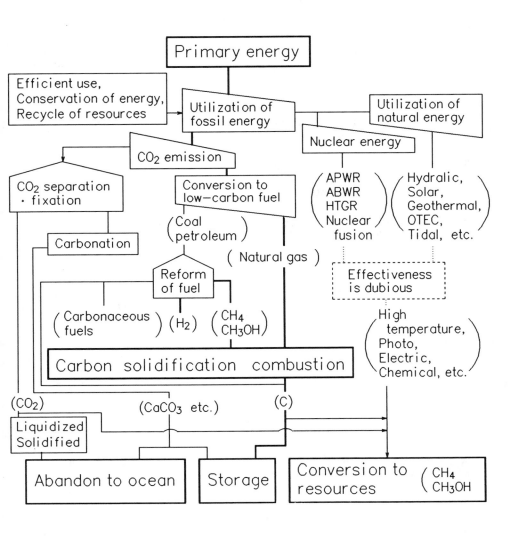

FIGURE 4.3.1 Countermeasures for $CO_2$ reduction

## Substantial reduction of $CO_2$ emission by the CSC

We consider here the combustion of a hydrocarbon fuel $C_mH_n$ on a per-mole-of-fuel basis. For a stoichiometric reaction the chemical equation is

$$C_mH_n + \left(m + \frac{n}{4}\right)(O_2 + 3.76N_2)$$
$$\rightarrow mCO_2 + \frac{n}{2}H_2O + \left(m + \frac{n}{4}\right)3.76N_2 \quad (4.3.1)$$

In the CSC, however, the hydrocarbon is not perfectly converted to $CO_2$ and $H_2O$; using fuel-rich mixture, only $(1-x)m$ mole of carbon contained in the fuel burns and converts to $CO_2$, and the residual is solidified as carbon. That is, the chemical equation is

$$C_mH_n + \left[m(1-x) + \frac{n}{4}\right](O_2 + 3.76N_2)$$
$$\rightarrow xmC + (1-x)mCO_2 + \frac{n}{2}H_2O + \left[m(1-x) + \frac{n}{4}\right]3.76N_2 \quad (4.3.2)$$

where $x$ is carbon solidification ratio.
The heating value for the reaction of Eq.(4.3.1) is

$$mQ_{CO_2} + \frac{n}{2}Q_{H_2O} - Q_{C_mH_n} \quad (4.3.3)$$

where, $Q_{CO_2}$, $Q_{H_2O}$ and $Q_{C_mH_n}$ are standard enthalpy of formation. On the other hand, the heating value for the reaction of Eq.(4.3.2) is

$$(1-x)mQ_{CO_2} + \frac{n}{2}Q_{H_2O} - Q_{C_mH_n} \quad (4.3.4)$$

Since the heating value for the CSC is smaller than that for the stoichiometric reaction, total fuel consumption in the CSC inevitably increases to compensate the energy which can be released from the solidified carbon. The increase in fuel consumption $y$ is given by the ratio:

$$y = \frac{mQ_{CO_2} + \frac{n}{2}Q_{H_2O} - Q_{C_mH_n}}{(1-x)mQ_{CO_2} + \frac{n}{2}Q_{H_2O} - Q_{C_mH_n}}$$

$$= \frac{m + \frac{n}{2}R_1 - R_2}{(1-x)m + \frac{n}{2}R_1 - R_2} \quad (4.3.5)$$

where $R_1 = Q_{H_2O}/Q_{CO_2}$, $R_2 = Q_{C_mH_n}/Q_{CO_2}$. Therefore, the substantial $CO_2$ emission by the CSC is expressed by

$$e = (1-x)y = \frac{(1-x)\left[m + \frac{n}{2}R_1 - R_2\right]}{(1-x)m + \frac{n}{2}R_1 - R_2} \quad (4.3.6)$$

FIGURES 4.3.2-4 show the relation between $x$, $y$ and $e$ for the case of methane ($CH_4$) and acetylene ($C_2H_2$). In Fig.4.3.2, the increase in fuel consumption $y$ varies nonlinearly against the carbon solidification ratio $x$, and reaches about 2 for methane and 2.7 for acetylene at the limit of perfect carbon solidification ($x = 1$). In Fig.4.3.3 the substantial $CO_2$ emission $e$ is less than unity for any value of $x$, which proves that the CSC satisfies a necessary condition for the countermeasure of $CO_2$ reduction. FIGURE 4.3.4 shows the relation between $y$ and $e$ as a function of $x$. It is found that to attain high $CO_2$ reduction ratio more

*Principle of combustion enhancement*
Since the combustion for fuel-rich mixture is essential in the CSC, the combustion enhancement on the basis of an excess enthalpy burning developed by Echigo et al.[4.3.6-8] is considered to be effective. A comparison between an ordinary gas-phase burning and excess enthalpy burning by using a porous radiative converter is illustrated in Figs.4.3.7(a) and (b) in terms of schematic temperature distributions along a flow direction. Under the condition of a normal flame as shown in Fig.4.3.7 (a), the temperature is raised by heat of combustion $h_o$ from an initial temperature $T_u$ to the final one $T_b$ (theoretical flame temperature) which is also the maximum temperature. On the other hand, in the case of Fig.4.3.7(b) the combustion gas enthalpy is effectively converted into thermal radiation emitted upward direction from the porous medium II, while the radiant energy is absorbed by the porous medium I and is converted into enthalpy increase in premixed gas. The energy recirculation produces a higher flame temperature $T_{max}$ than the theoretical one $T_b$, which means a high potential for the combustion enhancement relating to a low calorific fuel, namely, fuel-rich mixture in the present study.

*Apparatus*
On the basis of the excess enthalpy burning established by the radiative converter, a combustor was newly designed for the CSC as shown in Fig.4.3.8. Two porous media (made of ceramics) were installed in a quartz tube (i.d. 100mm) to form a combustion chamber; the porous medium I (nominal mesh length 1/30) was 50mm thickness and the porous medium II (nominal mesh length 1/6) was 25mm thickness. Methane-air mixture was

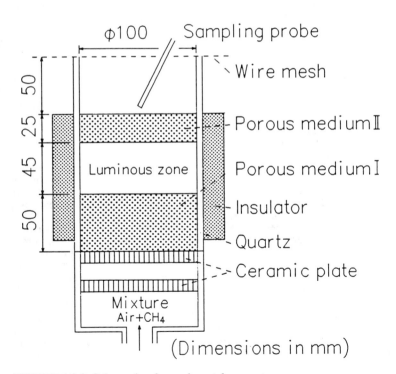

FIGURE 4.3.8 Schematic of experimental apparatus

supplied from the bottom of the quartz tube, and the one-dimensional flame was stabilized between the two porous media.

## Procedure

Flame was ignited at the open space above the wire mesh using a fuel-lean mixture. After the flame propagated to the region adjacent to the porous medium I, flammable limit was examined by changing the flow velocity $U$ and the equivalence ratio $\phi$. In the present experiments, only the fuel-rich side was tested because of the requirement of the soot formation. A chromel-almel thermocouple which has a radius $r$ of 0.05mm was used for the measurement of the gas temperature in the combustion chamber. In this case, along the flow direction the thermocouple was scanned quickly enough to be free from the effect of the soot deposition on the surface. The gas temperature $T_g$ was estimated from the measured temperature $T_t$ on the basis of the corrections for radiative heat losses and the time constant of the thermocouple, as follows.

$$\rho_t c_t \frac{dT_t}{d\tau} = \frac{2\alpha_m}{r}(T_g - T_t) - \frac{\varepsilon_t \varepsilon_p \sigma}{r}\left(2T_t^4 - T_p^4 - T_{gip}^4\right) \tag{4.3.10}$$

where $\rho_t$ and $c_t$ represent the density and specific heat of thermocouple, respectively. $\sigma T_p^4$ and $\sigma T_{gip}^4$ are radiation fluxes incident upon the upstream and the downstream side surface of the thermocouple, respectively, where $T_p$ is the temperature at the downstream end of the porous medium I and $T_{gip}$ is the temperature at the inflection point of the temperature profile in the downstream region from the maximum temperature. $\alpha_m$ is the mean heat transfer coefficient around the thermocouple. $\varepsilon_t$ and $\varepsilon_p$ are emissivities of the thermocouple and of the porous medium (and the side wall of combustor), respectively. A chromatographic analysis for burned gas was made with a water-cooled probe. Unburnt component was burned in the diffusion flame above the wire mesh.

## Results

The flammable limit for the present combustor is plotted in Fig.4.3.9 by open circles. Compared with the conventional flammable limit depicted by the solid line [4.3.5], the present technique largely extends the rich flammable limit up to 3.5; the luminous flame (yellow) was observed at about $\phi = 1.8$, and soot appeared $\phi \geq 2.0$. FIGURE 4.3.10 shows the typical temperature profile along the flow direction under the condition of the equivalence ratio of 3.0 and the mixture velocity of 2.4cm/s, where the abscissa denotes the distance from the downstream end of the porous medium I. The open circles indicate the gas phase temperatures $T_g$ estimated from the temperature $T_t$ (solid line) measured by the thermocouple. The mixture is preheated from the ambient temperature to about 600°C during the period when the gas flows through the porous medium I. The gas temperature is drastically increased by exothermic reaction in the combustion space near the porous medium I, and a luminous flame (yellow) is observed in the whole region of the downstream side from the exothermic reaction zone. On the other hand, in the luminous flame zone (yellow zone), the gas temperature is sharply decreased along the flow direction. This is because the large amount of energy is transferred by radiation from the luminous flame to the porous medium I; in the case of the luminous flame the emissive power for thermal radiation is as strong as that of the porous radiative converter, and further the endothermic reaction for the soot formation occurs in this region. The result of the chromatographic analysis for $U$=2.4 cm/s is shown in Fig.4.3.11. The carbon solidification ratio $x$ was obtained by calculating the difference between the amount of carbon in the burned gas and the initial carbon content in the fuel. For $\phi = 3.0$~$3.5$, $x$ attains about 20%, and the solidified carbon was found to attach on the surface of the porous medium II. The results are remarkable considering that in the conventional methane combustion soot was virtually nonexistent. Thus, we conclude that the way to the CSC has been opened by the radiation-controlled combustion with porous media.

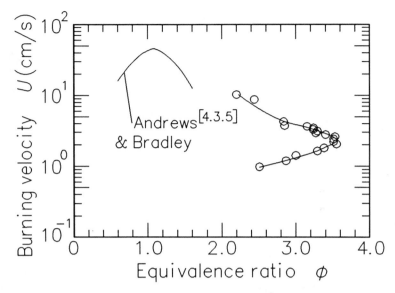

FIGURE 4.3.9 Flammable limit of methane combustion

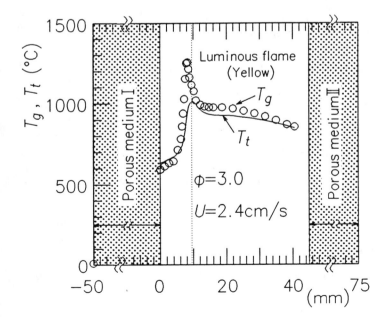

FIGURE 4.3.10 Temperature profile

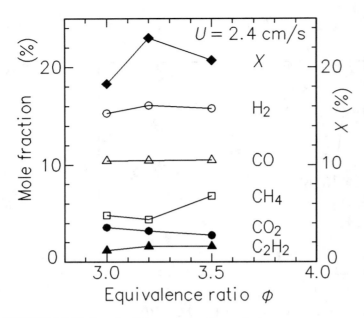

FIGURE 4.3.11 Result of chromatographic analysis

**Concluding remarks**

The mass consumption of the fossil fuels had commenced in the 1960's owing to the expediency of the petroleum and has enhanced the anthropogenic activities to a great extent. The conversion of fuel from coal to petroleum is called "an energy revolution", which has contributed remarkably to construct the highly industrialized and civilized countries. In contrast to the coal, one of the pronounced features of the petroleum is to be burned without the ash, which enhanced the mass consumption of energy. However, figuratively speaking, $CO_2$ should be regarded to be "gaseous ash" which brings about the global warming.

The basic idea of the CSC is to produce the "quasi-ashes" of soots in solid phase even for the combustion of the gaseous fuels. To this end the combustion technologies have borne the new aspects of science and engineering for building up a novel energetics. The advanced combustion technology may provide, for instances, the precise control methodology to produce the soots with a lower content ratio of hydrogen, the sophisticated energy system for hydrocarbon-watervapor reforming and so on. While we experienced, as a similar problem, the injurious gaseous emissions of $SO_x$ and $NO_x$ in the past, the problem of $CO_2$ emission into atmosphere is different in quality and by far the most difficult to solve, never experienced so far.

**References**

[4.3.1] Revelle, R., Carbon Dioxide and World Climate, *Scientific American*, vol.247, no.2, pp.33-41, 1982.

[4.3.2] Steinberg, M., and Cheng, H.C., Modern and Prospective Technologies for Hydrogen Production from Fossil Fuels, *Proc. 7th World Hydrogen Energy Conf.*, 1988.

[4.3.3] Seifritz, W., Methanol as a Link between Primary Fossil Energy and a Carbon Dioxide-free Energy System, *Proc. 7th World Hydrogen Energy Conf.*, 1988.

[4.3.4] Steinberg, M., A Process Concept for Utilizing Fossil Fuel Resources with Reduced $CO_2$ Emission, *Int'l Conf. on Coal Science-IEA*, vol.2, pp.1059-1062, 1989.

[4.3.5] Andrews, G.E., and Bradley, D., The Burning Velocity of Methane-Air Mixtures, *Combustion and Flame*, vol.19, no.2, pp.275-288, 1972.

[4.3.6] Echigo, R., Effective Energy Conversion Method between Gas Enthalpy and Thermal Radiation and Application to Industrial Furnaces, *Proc. 7th Int. Heat Transfer Conf.*, Munchen, vol.6, pp.361-366, 1982.

[4.3.7] Echigo, R., Kurusu, M., Ichimiya, K., and Yoshizawa, Y., Combustion Augmentation of Extremely Low Calorific Gases (Application of the Effective Energy Conversion Method from Gas Enthalpy to Thermal Radiation), *Proc. 1983 ASME-JSME Thermal Engineering Joint Conf.*, Honolulu, vol.IV, pp.99-103, 1983.

[4.3.8] Yoshizawa, Y., Sasaki, K., and Echigo, R., Analytical Study of the Structure of Radiation Controlled Flame, *Int.J. Heat Mass Transfer*, vol.31, no.2, pp.311-319, 1988.

# 5. HYDROGEN FUEL

## 5.1 PRODUCTION, STORAGE, TRANSPORTATION AND UTILIZATION OF HYDROGEN

**Etsuo Akiba**
National Chemical Laboratory for Industry
Agency of Industrial Science and Technology
Ministry of International Trade and Industry
1-1 Higashi, Tsukuba, Ibaraki 305, Japan

### 5.1.1 Introduction

Hydrogen is produced from water and it can be used for fuel. Water is formed again by combustion of hydrogen with oxygen in the air. Hydrogen is an ideal fuel because hydrogen itself and gases formed by the combustion of hydrogen are not greenhouse and ozone layer damaging gases. Therefore, hydrogen is the most environmental friendly fuel that we have ever had.

Electric power and fuel are major secondary energies. Secondary energy is a category of energy in which its final form of energy can be consumed. Hydrogen seems to be a promising future secondary energy. It is expected that research and development on hydrogen fuel will be conducted much more actively from now on.

Hydrogen gas does not naturally exist. Therefore, hydrogen must be produced from hydrogen containing compounds such as water and hydrocarbons by adding energy. At present, hydrogen is produced in large scale as a raw material for the synthesis of ammonia, methanol and other chemicals but not for fuel. In other words, hydrogen fuel has not been realized but will be actualized in the near future. In this chapter hydrogen will be discussed as fuel which will be used for aircraft, space application, power generation, combustion, etc. Especially, production of hydrogen is a very important technology for achieving hydrogen energy systems. Storage, transportation and utilization of hydrogen fuel will also be discussed.

### 5.1.2 Hydrogen gas

Hydrogen is the lightest element. At room temperature, hydrogen is a colorless gas existing as a $H_2$ molecule. However, hydrogen exists as a component of compounds like hydrocarbons and water not but as free gas on earth. Hydrogen is the 9th richest element on earth. The characteristics of hydrogen gas are listed in Table 5.1.1. The density of hydrogen is the smallest, 83.8g/m$^3$, at 20°C. It means that the energy density per weight is high but that the density per volume is not expected to be high. Hydrogen has very good heat conductivity so that hydrogen is filled inside of an electric power generator in order to remove the heat produced inside of the generator and to decrease the energy loss due to viscous drag.

Hydrogen is liquified at -239.9°C and 12.8atms. These values are called its critical temperature and critical pressure, respectively. The hydrogen molecule consists of two hydrogen atoms. Each atom has a nuclear spin and therefore, there are two kinds of hydrogen molecules; ortho hydrogen with different spins and para hydrogen with a parallel spin. The ratio of ortho-para hydrogen is 3:1 at room temperature but para hydrogen is more stable at lower temperatures. Therefore, when hydrogen gas is liquified ortho hydrogen should be transformed to para using a catalyst, otherwise the heat of transformation from liquified ortho hydrogen to para hydrogen will cause vaporization of the liquid hydrogen.

Hydrogen has three isotopes; hydrogen(protium) H, deuterium D and tritium T. They are identical in chemical properties but different in physical properties. The natural abundance of protium is 99.985% and that of deuterium is 0.015%. Tritium is radioactive and does not exist naturally. The fusion reaction between deuterium and tritium is used in a fusion reactor. Recently, a new phenomena called "cold fusion" was found [5.1.1]. Neutron and "excess heat" were detected when heavy water($D_2O$) was electrolyzed using a Pd cathode. In this article, we will discuss hydrogen as a fuel and a media for energy transportation. Therefore, the isotopes of hydrogen should not be considered because the nature of the fuel is determined by the chemical properties of hydrogen.

The reaction between hydrogen and oxygen, combustion or explosion of hydrogen, is one of the well known chemical reactions. This reaction is also accepted as a dangerous reaction. People still remember the explosion of the "Hindenburg" in 1939. Actually the burning ratio of hydrogen to air is very wide; 4-75Vol%. The explosion mixture ratio is also very wide and its lowest end is very low. The ignition energy of hydrogen is about 1/10 of methane or gasoline. The flame of hydrogen burning runs very fast and is colorless. Hydrogen gas has a high diffusion rate, which is three time higher than methane. Therefore, it is reasonable that people believe that hydrogen is a very dangerous gas. In addition, hydrogen embrittlement is a problem to solve.

However, recent progress in gas sensor technology makes detection of hydrogen easier. Research on hydrogen embrittlement has established an empirical rule from the intensive investigation of many accidents caused by hydrogen embrittlement [5.1.2]. Hydrogen is not self explosive, therefore, hydrogen should be handled in essentially an identical manner as other flammable gases. In conclusion, hydrogen is not a dangerous gas if it is handled properly.


5.1.3 Production of hydrogen

In this section, the production methods of hydrogen will be discussed in terms of kinds of energy added the system for decomposition of water or other compounds. These are heat, electric power, light(solar energy) and biological systems.

### 5.1.3.1 Hydrogen production by heat

(A). Hydrogen from hydrocarbons

The most plausible methods for hydrogen production from hydrocarbons are the steam reforming process and the partial oxidation process. Light hydrocarbons such as naphtha and methane(natural gas) are decomposed by the steam reforming process while much heavier hydrocarbons are used for hydrogen production using the partial oxidation process.

At present the steam reforming of natural gas(methane) is the most economic large scale hydrogen production method. The reactions are as follows;

$$CH_4 + H_2O \rightarrow CO + 3H_2 \quad (5.1.1)$$
$$CO + H_2O \rightarrow CO_2 + H_2 \quad (5.1.2)$$

The reaction (5.1.1) is an endothermic reaction, therefore, heat for the reaction must be supplied from outside of the system. The reaction temperature is normally 800°C and reaction pressure is 7-35atms. The gas mixture produced in the reaction (5.1.1) is often called *syn* gas because this mixture is a raw material for the synthesis of many useful hydrocarbons. The reaction of (5.1.2) is the "water shift reaction". The reaction is shown in an equation but the actual process is a two step reaction. The reaction temperatures are 400°C and 200°C. After the reaction (5.1.2), CO and $CO_2$ are removed and pure hydrogen is obtained. LNG and naphtha are also used for the raw materials in the steam reforming process.

Hydrogen is also produced by the partial oxidation of hydrocarbons using steam and pure oxygen. The raw materials for partial oxidation process are not only light hydrocarbons which can be used for steam reforming but also heavy hydrocarbons such as heavy oil, crude oil and tar. The reactions are shown as follows;

$$C_mH_n + mH_2O \rightarrow mCO + (m+n/2)H_2 \quad (5.1.3)$$
$$CO + H_2O \rightarrow CO_2 + H_2 \quad (5.1.4)$$

The plant for this process is operated at 1100-1400°C and at more than 30atms. The plant for the partial oxidation process is much simpler than that for steam reforming. However, the disadvantage of the partial oxidation process over steam reforming is that it needs pure oxygen.

Methanol is easily decomposed to form carbon monoxide and hydrogen.

$$CH_3OH \rightarrow CO + 2H_2 \quad (5.1.5)$$

Formed CO can be converted to hydrogen and carbon dioxide again by the water gas shift reaction.

$$CO + H_2O \rightarrow CO_2 + H_2 \quad (5.1.6)$$

The reaction (5.1.5) is endothermic. Therefore, heat of the reaction should be supplied from outside of the system. This reaction can proceed in both directions without difficulty. Therefore, methanol is used for hydrogen transportation on a relatively small scale. Methanol is synthesized in a chemical plant and delivered to the place of consumption. Hydrogen is obtained there by decomposition of methanol and the water gas shift reaction. If we compared this method to the steam reforming process, the cost for construction of the plant is lower for the methanol process but the cost for raw material is higher than that of the steam reforming. Therefore, the methanol process has an advantage in small scale transportation.

(B) Hydrogen production by thermochemical reactions

The combination of chemical reactions using alkali earth metals, iron, sulfur and halogen, etc. provides a method for the splitting of water into hydrogen and oxygen under relatively moderate conditions. Water is a very stable compound. It must be heated to more than 3000°C if it is to be decomposed purely by heat. Thermochemical hydrogen production was first proposed by Funk and Reistroam [5.1.3] in 1966. They proposed that the heat of decomposition of water would be supplied stepwise using several reactions and that the maximum reaction temperature can be kept at 1000°C. Therefore, the high temperature gas reactor which can supply heat at about 1000°C will be used as the heat source for this process. In the thermochemical water production process, a reaction cycle is formed by several chemical reactions. Water and heat of the reaction are supplied externally but the chemicals which are part of the reaction cycle are circulated inside the plant. The final products are hydrogen and oxygen. This process is always compared to electrolysis in energy efficiency. The target efficiency of thermochemical reactions is 30% which is the product of efficiencies of power generation and electrolysis. At first, the thermochemical process was expected to have a much higher efficiency than electrolysis, because this is a one step energy conversion from thermal energy to hydrogen.

After this process was proposed, many organizations started R & D on this method because of its higher efficiency. Larger scale and newer processes were expected. Some results of the efforts are shown in Table 5.1.2. However, the pure thermochemical hydrogen production process was recognized to have some technological disadvantages as follows, 1)the reactions which constitute the cycles could proceed at least 90% or more, 2)the reaction products of each reaction should be easily separated from the mixture of the chemicals, 3)a large amount of chemicals should be circulated in order to decompose only a small amount of water, 4)almost all candidate chemicals such as sulfur, halogen, alkali and alkali earth meals are very corrosive and the plant should be built of special materials, and 5)if solid state chemicals are used in the thermochemical process, handling of the chemicals will be much more difficult.

At this moment, this process has been developed by some public organizations and universities. The R & D program for this process should be a long term project, because the thermochemical process is promising in principle but it has many technological difficulties at present.

5.1.3.2 Hydrogen production by electrolysis

The cost of hydrogen which is produced by electrolysis depends very much upon the cost of electricity. Actually, the cost for power is about two thirds of the total cost [5.1.4]. The technology for water electrolysis is very simple and the costs for construction of the plants are low. Electrolysis has another disadvantage when compared to the other technology. That is the difficulty in scaling up because the electrodes and electrolytic cells have limitations in size. In addition, the plants which require hydrogen such as that for ammonia and methanol, were significantly scaled up but the size of the electrolysis hydrogen production system did not match those plants. In the countries which import energy resources such as Japan and Germany, the cost of hydrogen produced by electrolysis is quite high as shown in Fig. 5.1.1 [5.1.5]. However, those countries which have rich hydro power such as Canada and Brazil have the potential to produce low cost hydrogen from hydro power using electrolysis. The relation between power and hydrogen costs is shown in Fig. 5.1.2 [5.1.4]. In conventional electrolysis cells, nickel coated or nickel alloy electrodes and KOH concentrated electrolyte solution (25-35%) are normally used.

In order to overcome the cost problem, the efficiency of electrolysis has been improved under the Sunshine project in Japan. The efficiency of the conventional electrolysis is 60-70%. Electrolysis was conducted under high pressure and high temperature conditions. The electrodes, the electrolyte and the design of whole cell were also improved. This R &D project finished in 1984. The obtained performance was as follows; hydrogen uptake, 20$Nm^3$/h; cell temperature, 90-120°C; cell pressure, 20 atms; and cell voltage, 1.8V at 20-50 A/$dm^2$. However, the cost of hydrogen made by this improved technology is still high in comparison with other hydrogen production methods in Japan [5.1.6]

One of new technologies to improve the electrolysis of hydrogen is solid state polymer electrolyte (SPE) electrolysis. In this system a thin polymer film plays the role of both the electrolyte and a separator. On both sides of the polymer film catalytic electrodes are attached as shown in Fig. 5.1.3. Therefore, pure water is supplied to the cell instead of concentrated alkali solution such as KOH. The advantages of this method are high current density, lower overvoltage and compactness of the system.

A precise cost estimation of this method has not been carried out. However, a rough estimation suggests that polymer film and the catalytic electrodes are quite expensive. At this moment, the SPE electrolysis system is very suitable for the small scale usage of hydrogen. In Japan, a hydrogen supplier using this technology was put into the market for laboratory scale hydrogen users (Fig. 5.1.4).

The efficiency of electrolysis becomes higher at higher temperature. Therefore, high temperature electrolysis technology is being investigated. One of the promising methods is electrolysis at about 1000°C using solid state electrolyte cells. In other words, the reaction of this method is the reverse reaction of solid state oxide fuel cells(SOFC). The decomposition of water requires an exothermic reaction, therefore, the cell should be kept at higher temperature by thermal energy from the outside. For this purpose, the high temperature atomic pile will supply energy to the electrolytic cells. In technical aspects, to build the solid state electrolysis system is considered to be too difficult than to make the solid state oxide fuel cell. However, the materials for

Table 5.1.2 Reactions of Thermochemical Cycles for thermochemical hydrogen production

Fe-Cl cycle

$3FeCl_2 + 4H_2O \rightarrow Fe_3O_4 + 6HCl + H_2$
$Fe_3O_4 + 3/2Cl_2 \rightarrow 3FeCl_3 + 3H_2O + 1/2O_2$
$3FeCl \rightarrow 3FeCl_2 + 1/2Cl_2$

UT-3 cycle

$CaBr_2 + H_2O \rightarrow CaO + 2HBr$
$CaO + Br_2 \rightarrow CaBr_2 + 1/2O_2$
$Fe_3O_4 + 8HBr \rightarrow 3FeBr_2 + 4H_2O + Br_2$
$3FeBr_2 + 4H_2O \rightarrow Fe_3O_4 + 6HBr + H_2$

GA cycle

$xI_2 + SO_2 + 2H_2O \rightarrow H_2SO_4 + 2HI_x$
$2HI_x \rightarrow H_2 + xI_2$
$H_2SO_4 \rightarrow H_2O + SO_2 + 1/2O_2$

NIS cycle

$2H_2O + I_2 + SO_2 \rightarrow 2HI + H_2SO_4$
$2HI + H_2SO_4 + 2Ni \rightarrow NiI_2 + NiSO_4 + 2H_2$
$NiI_2 \rightarrow Ni + I_2$
$Ni_2SO_4 \rightarrow NiO + SO_3$
$SO_3 \rightarrow SO_2 + 1/2O_2$
$NiO + H_2 \rightarrow Ni + H_2O$

Mg-S-I cycle

$I_2 + SO_2 + 2H_2O \rightarrow H_2SO_4 + 2HI$
$2MgO + H_2SO_4 + 2HI \rightarrow MgSO_4 + MgI_2$
$MgI_2 \rightarrow MgO + 2HI + nH_2O$
$MgSO_4 \rightarrow MgO + SO_3$
$SO_3 \rightarrow SO_2 + 1/2O_2$
$2HI \rightarrow H_2 + I_2$

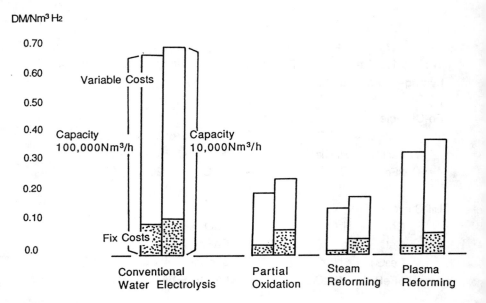

Fig. 5.1.1 Production Costs for H2 in Large-scale Processes
H2 quality: 99.9Vol%, H2 Pressure: 25bar, Price Level 1985 [5.1.5]
(Reprinted with permission from *Hydrogen Energy Progress VI*, eds. T. N. Veziroglu, N. Getoef and P. Weinzierl, Copyright 1986, Pergamon press PLC.)

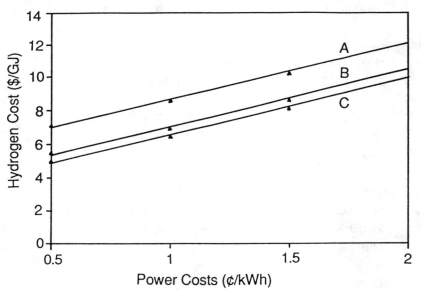

Fig. 5.1.2 The Relation between Power and Hydrogen Costs [5.1.4] (Plant Capacity A 660Ml H2/day, 11MW; B 2350Ml H2/day, 37.5MW; C 6600Ml H2/day, 110MW) (Reprinted with permission from *Hydrogen Energy Progress VII*, eds. T. N. Veziroglu and A. N. Protsenko, Copyright 1988, Pergamon Press PLC.)

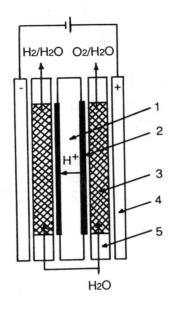

Fig. 5.1.3 Solid State Polymer Electrode (SPE)

1 SPE; 2 Catalytic Electrodes; 3 Collectors, 4. Electrodes; 5 Gasket

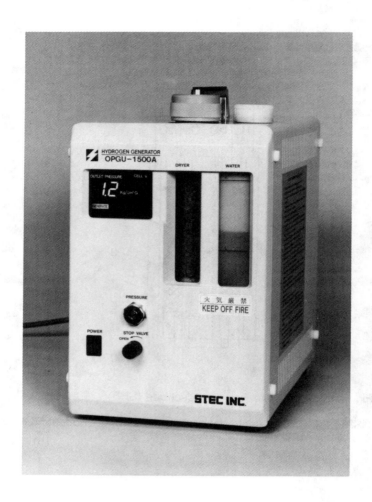

Fig. 5.1.4 Small Scale Hydrogen Production Unit Using SPE Electrodes

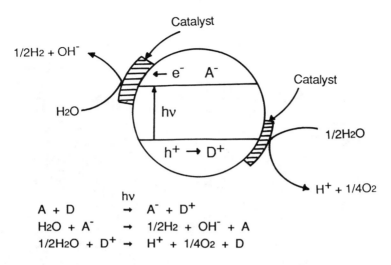

Fig. 5.1.5 Schematic Drawing of Photocatalytic Hydrogen Production Process

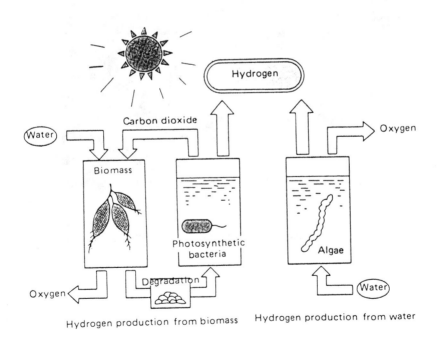

Fig. 5.1.6 Hydrogen Production by Biological Systems

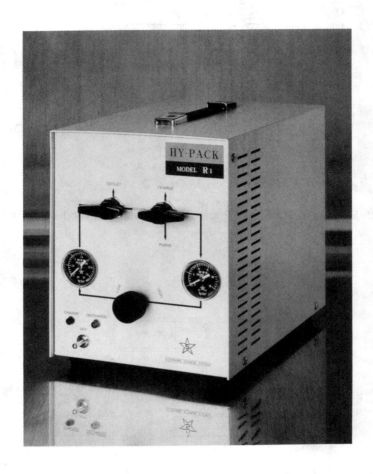

Fig. 5.1.7 Hydrogen Container Using Metal Hydrides (Suzuki Shokan)

hydrogen is supplied economically, the alkali solution fuel cells will be dominant.

### 5.1.5.3 Automobiles

One of the most important utilization of hydrogen is for automobiles because hydrogen is the fuel in the future. Details will be discussed in another chapter. Therefore, the importance of the automobile application is shown only briefly here. The hydrogen fueled automobile has a merit at present. Automobiles today are one of the most $CO_2$ productive applications so they should be converted to less or no $CO_2$ producing ones. There are two types of hydrogen storage; liquid hydrogen and metal hydrides. The former one will be described later. The major difference in both systems is hydrogen pressure from the storage tanks [5.1.12]. As a result, the fuel supply system and the design of the engine will be different.

### 5.1.5.4 Other application

Hydrogen can be used for town gas. However, NOx formation and back fire are problems to solve. The flame of hydrogen burning is colorless, then coloring agents must be attached to the burner for family use. Catalytic combustion is ideal for hydrogen utilization in private housings. The R & D for this application has been carried out and the results suggests that technology has reached the application level.

Hydrogen gas is now used for cutting steel sheets and for working precious metals and artificial jewels. Hydrogen flame plays an important role in triggering the reaction between materials and oxygen.

### 5.1.6 Hydrogen fuel in the future

Hydrogen has two faces; raw materials and secondary energy (fuel). As a secondary energy, hydrogen is the only fuel that do not form greenhouse gases. Recently, new technology such as HST and the fuel cells which need hydrogen fuel is being investigated and developed to a certain level. Therefore, hydrogen fuel is being discussed much more realistically than had been done during the oil crisis about a decade ago.

In addition, there are large scale projects that involves hydrogen transportation between continents. The Euro-Quebec Hydro-Hydrogen project is a good example [5.1.13]. That is an investigation on hydro-hydrogen based on a clean energy system of a 100MW level. The concept is shown in Fig. 5.1.9. Hydrogen is produced by electrolysis generated by hydro power in Quebec, Canada. Hydrogen is transported in the form of liquid hydrogen or methylcyclohexane (MCH) to Europe. There, hydrogen will be used for power and/or heat generation, vehicle and aircraft, and a component of town gas. This feasibility study is financed by the commission of the European Communities, the Quebec provincial government and associated industries.

MCH can contain 6 rechargeable hydrogen atoms in a molecule. Hydrogen transport reaction of MCH is shown as follows.

$$\text{methylcyclohexane} \leftrightarrow \text{toluene} + 3H_2 \quad (5.1.11)$$

The total amount of hydro power generation was about 2,000TWh/year in 1988 [5.1.14]. The technically exploitable hydro power lies in a range of 15,000 to 19,000TWh/year. This amount is about a quarter of the worldwide primary energy consumption. It should be noted that hydro power does not produce any pollutants. However, most locations which have the potential for large amounts of hydro power are far from the place of energy consumption. That is the reason that hydrogen is used as the cleanest and storable energy carrier.

In Japan, similar feasibility studies have been carried out. In the Euro-Quebec project, hydrogen is produced by hydro power but hydrogen can be produced by the steam reforming process from natural gas at the site of the gas well. Formed $CO_2$ is treated at the site(one of the suitable $CO_2$ treating methods is described in 4.1) and hydrogen is then transported to the center of energy consumption [5.1.14].

The Japanese government investigated the feasibility of "clean energy transportation". The concept is that hydrogen which is produced by renewable energy such as hydro and solar outside Japan is transported to Japan in several forms. Those forms are liquid hydrogen, methanol, ammonia, cyclohexane, urea, formic acid and metal hydrides [5.1.15]. The report concludes that the cost of electricity produced by hydrogen transported as previously mentioned lies in the range of 15-20¥/kWh which shows the possibility of the already mentioned technology.

The trend in hydrogen fuel R & D indicates that hydrogen is being accepted as one of the promising fuels of the future. The investigations have become more realistic than before. However, the most important problem to solve is not in technology but in public acceptance. Generally speaking, people still feel that hydrogen is very dangerous. An effort to publicize the merits and proper handling of hydrogen energy is needed. Very recently, new small size rechargeable batteries are being marketed, in which hydrogen is used for the storage of electricity. Such technology makes a new step for the acceptance of hydrogen energy.

## References

[5.1.1] M. Freischmann, S. Pons and M. Hawkins, *J. Electroanal*, **261**, 301 (1989); **263**, 187 (1989): S. E. Jones, E. P. Palmer, J. B. Czirr, D. L. Decker, G. L. Jensen, J. M. Thorne, S. F. Taylor and J. Rafelski, *Nature* (London), **338**, 737 (1989).

[5.1.2] for example, J. G. Morris, in *Hydrogen Energy Progress*, eds. T. N. Veziroglu, K. Fueki and T. Ohta, pp 1491-1508, Pergamon, New York, 1980.;H. G. Nelson, in *Hydrogen Energy Progress V*, eds. T. N. Veziroglu and J. Bryan Taylor, pp 1841-854, Pergamon, New York, 1984.

[5.1.3] J. E. Funk and R. M. Reinstrom, *Ind. Eng. Chem. Process Des. Dev.*, **5**, 336 (1966)

[5.1.4] J. O'M. Bockris and J. C. Wass, in *Hydrogen Energy Progress VII*, eds. T. N. Veziroglu and A. N. Protsenko, pp 101-151, Pergamon, New York, 1988.

[5.1.5] G. Kaske, L. Kerker and R. Müller, in *Hydrogen Energy Progress VI*, eds. T. N. Veziroglu, N. Getoef and P. Weinzierl, pp 185-196, Pergamon, New York, 1986.

[5.1.6] Sunshine Project Promotion Office, *Outline of Sunshine Project*, 1988, (1988).

[5.1.7] A. Fujishima and K. Honda, *Nature* (London), **238**, 37 (1972).

[5.1.8] T. A. Vardapetyan and A. V. Vartanyan, in *Photocatalytic Conversion of Solar Energy*, eds. V. N. Parmon and K. I. Zamaraev, Nauka, Novosibirsk, 1990.

[5.1.9] V. N. Parmon, in *Hydrogen Energy Progress VIII*, eds. T. N. Veziroglu and P. K. Takahashi, pp 801-813, Pergamon, New York, 1990.

[5.1.10] H. Wenzl, *Int. Metals Rev.*, **27**, 140 (1982).

[5.1.11] G. D. Brewer, *Astrounautics & Aeronautics*, May 1974.

[5.1.12] J. Hama, Y. Uchiyama and Y. Kawaguchi, *SAE paper*, 880036, (1988).

[5.1.13] R. Wurster and A. Malo, in *Hydrogen Energy Progress VIII*, eds. T. N. Veziroglu and P. K. Takahashi, pp 59-70, Pergamon, New York, 1990.

[5.1.14] Committee for Research on Resources of Liquid Hydrogen and Rare Gases, *Report of Research on Resources of Liquid Hydrogen and Rare Gases*, 1988. (in Japanese).

[5.1.15] *Report of Overseas Transmission of Clean Energy, 1989*, Organization for Engineering Promotion, 1990. (in Japanese).

# 5.2 HYDROGEN FUELED ENGINES FOR VEHICLE

Shoichi Furuhama
Musashi Institute of Technology
1-28-1 Tamazutsumi
Setagaya-ku Tokyo 158, Japan

INTRODUCTION

Recently, increase of $CO_2$ in the air has been considered as a serious problem, and its reduction has been urged. Amounts of $CO_2$ production per unit rate of heat release are compared in Table 5.2.1, where it is shown that relatively small difference exits among fossil fuels in terms of $CO_2$ production, but that hydrogen combustion produces absolutely no $CO_2$. There are still a number of difficulties in order to make hydrogen widely used as a stndard fuel. Some of them are economical production, effective storage, feasible applications, and reliable safety measures.

The author has been working on hydrogen fueled automobile engines and believes that if feasibility of the hydrogen engine is fully demonstrated with respect to its performance, reliability and man-machine interface, other difficulties pointed out above may also be overcome earlier than anticipated. In such an expectation, the auther would like to describe a brief history of the hydrogen engine reserch and development conducted at Musashi I. T. Engin Laboratory.

5.2.1. Properties of Hydrogen ( Comparison with Gasoline or $C_8H_{18}$)

Hydrogen properties which are closely related to the engine perfurmance are compared with those of gasoline in Table 5.2.2. The density ratio of gasoline vapor and gaseous hydrogen is 1 : 0.018, while their liquid density ratio is about 1 : 0.1. Although the weight ratio of hydrogen over air under stoichiometric reaction is only 45 % of the corresponding value of gasoline, the hydrogen volume ratio under the same condition becomes nearly 30 %, while the gasoline value is only 1.7 %. As a result, Calorific value per unit mol of fuel-air mixture for hydrogen combustion is 85 % of that for gasoline combustion, which means that the hydrogen engine can extract 85 % as large power output as the gasoline engine if pre-mixture combustion method as shown in Figure 5.2.1-(b) is selected. On the other hand,

Table 5.2.1  $CO_2$ [kg/(Q/kcal)] as a combustion product of various fuels.

| Fuel | Typical molecular | Calorific value | $CO_2$ | Comparison with gasoline |
|---|---|---|---|---|
| Coal | C | 8100 | 3.54 | 1.56 |
| Diesel fuel | $C_{16}H_{34}$ | 10590 | 2.28 | 1.01 |
| Gasoline | $C_8H_{18}$ | 10630 | 2.27 | 1.00 |
| Methanol | $CH_3OH$ | 4770 | 2.26 | 0.99 |
| Natural gas | $CH_4$ | 11930 | 1.80 | 0.79 |
| Hydrogen | $H_2$ | 28700 | 0 | 0 |

since Calorific value per unit mol of air is 20 % higher for hydrogen than for gasoline, the hydrogen engine power output turns out to be 120 % of that for the gasoline engine if hydrogen is injected after air alone is taken into the engine and intake valve is closed, as shown in Figure 5.2.1-(d).

It is also indicated in Table 5.2.2 that very lean hydrogen-air mixture, which corresponds to 0.14 of the stoichiometric mixing ratio, can be ignited by spark with much smaller energy than in the gasoline case. As a result of such feature, power output of the hydrogen engine can be controlled by means of changing hydrogen flow rate only, while in the case of the gasoline engine, air flow rate as well as gasoline flow rate should be changed. However, the fact that self ignition temperature for hydrogen is higher than that for gas oil makes impossible the compression ignition. The data in Table 5.2.2 shows that adiabatic flame temperature under constant pressure combustion is slightly higher for hydrogen than for gasoline.

Although diffusion coefficient for hydrogen is eight times as large as that for gasoline, the coefficient is not large enough to promote mixture formation. However, this large diffusion coefficient implies swift diffusion of hydrogen into air and, as a result, high safety even if it leaks. Quench distance is the minimum gap at which the flame can exist, and according to the data, hydrogen flame tends to propagate through a narrow gap.

Hydrogen can be ignited very easily by high temperature spot such as spark. Although this nature makes lean mixture combustion possible, it is also a primary reason why hydrogen is considered to be dangerous. In addition, this nature causes backfire, by which the mixture of hydrogen and air taken into a cylinder is ignited before the intake valve is closed and explosively burns inside of the intake pipes. The backfire is regarded as pre-ignition at the very early stage, but its mechanism has not been clarified yet. Although various countermeasures against the backfire have been proposed, none of them can prevent it completely as long as hydrogen is mixed with air outside of the cylinder before entering the cylinder. The backfire usually occurs when the excess air ratio $\lambda$ is $1.5 \sim 2.0$ or less, thus the pre-mixture hydrogen engine can not be operated under richer mixture conditions than $\lambda = 1.5 \sim 2.0$. Since the hydrogen engine power output at $\lambda = 1.0$ is 85 % of that for the gasoline engine, this operation restriction causes its maximum power output to be only 50 % of the gasoline engine maximum.

5.2.2. Carriage of Hydrogen

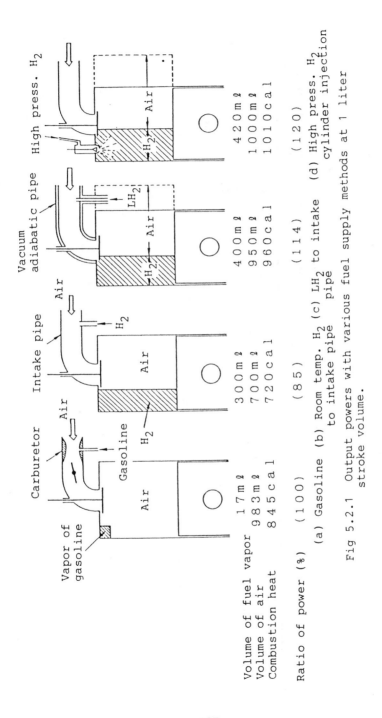

Fig 5.2.1  Output powers with various fuel supply methods at 1 liter stroke volume.

(a) Gasoline  (b) Room temp. $H_2$ to intake pipe  (c) $LH_2$ to intake pipe  (d) High press. $H_2$ cylinder injection

Table 5.2.2  Comparison of properties between hydrogen and gasoline.

| | Hydrogen $H_2$ or $LH_2$ | Gasoline (Mean value) | Ratio of $H_2$/Gasoline |
|---|---|---|---|
| Specific gravity (gas:50°C,1atm) (liquid) | $0.076 \times 10^{-3}$ 0.071 | $4.23 \times 10^{-3}$ 0.735 | 0.018 0.097 |
| Calorific value  kcal/kg | 28700 | 10650 | 2.70 |
| Fuel/air ratio (weight) at $\lambda=1.0$ (volume) | 0.0284 0.296 | 0.063 0.017 | 0.45 17.4 |
| Calorific value of mixture kcal/mol | 17100 | 20200 | 0.85 |
| Calorific value per air vol. kcal/mol | 24300 | 20500 | 1.19 |
| Limit of inflammability (vol. F/A) | 0.04 | 0.012 | |
| Self ignition temperature (at 1atm) °C | 580 | 500, (340※) | |
| Minimum ignition energy mJ | 0.02 | 0.25 | 0.08 |
| Adiabatic flame temp. °C | 2110 | 2000 | |
| Coefficient of diffusion $cm^2/s$ | 0.63 | 0.08 | 8 |
| Min. quench distance cm | 0.06 | | |

※ Gas oil

Gaseous hydrogen has the smallest density, which is 1/14.5 of air density, therefore it seems impossible to use hydrogen as transportation fuel in gaseous state under ambient pressure. As shown in Figure 5.2.2, although the weight ratio of hydrogen and gasoline at the same Calorific value is about one-third (0.37), the volume of hydrogen gas at room temperature is 3000 times as large as gasoline volume, and even liquid hydrogen at 20 K has 3.85 times as large volume. From this viewpoint, methods of hydrogen carriage is regarded as one of the most important and the most difficult issues for hydrogen fueled vehicles. The following three methods have been considered to be promising candidates and tested.
(a) Metal-Hydride ( MH )
(b) High Pressure ( 150 ~ 200 atm) Tank ( HP )
(c) Liquid Hydrogen ( $LH_2$ )
Developments of MH for automobile use have been conducted at various institutes such as Benz Co. However, weight of MH is significantly large and is as large as HP, as shown in Table 5.2.3. It should be also pointed out that hydrogen taken out of MH has such low pressure that hydrogen injection into the cylinder can not be realized. This disadvantage is serious for engine application because without the injection method, backfire can not be prevented as above-mentioned, and the engine power remains as low as 50 % of the gasoline engine. Therefore, unless MH is improved to be much lighter and to be capable of high pressure hydrogen discharge, application of MH to automobiles is not feasible.

High Pressure Tank (HP) method has also difficulty in its weight. Although high pressure hydrogen can be discharged when the tank contains sufficient hydrogen, pressure inside of the tank gradually decreases, and when the pressure becomes lower than the injection pressure, hydrogen in the tank can not be taken out any more. Many efforts have been made to develop small weight HP such as the one made of aluminum and strengthened by carbon wire around the tank. The high pressure tank shown in Figure 5.2.3, which was recently proposed by Krepec et al.[5.2.1] at Concordia University in Canada, is noteworthy because its structure is simple but effective. Stored hydrogen in this tank is under the condition at 100 K and 200 atm . Therefore, its thermal insulation of tank is easier than the $LH_2$ of 20 K, and storage mass of hydrogen can be triple as much as that at 300 K. In addition, effectiveness of the tank, which is defined as a ratio of usable amount to full storage, becomes higher than room temperature HP. For example, if hydrogen is injected at 100 atm , hydrogen can be taken out of this tank until pressure and temperature inside of the tank reaches 100 atm and 300 K, respectively. And, in this case, its effectiveness becomes 5 times as compared to a conventional high pressure hydrogen container. One conceivable problem is that if

Table 5.2.3 Comparison of fuel storage weights correspond to gasoline 30 liter.

| Fuel tank | Content | | Tank weight (kg) | Total weight (kg) |
| --- | --- | --- | --- | --- |
| | Volume (l) | Weight (kg) | | |
| Gasoline | 30 | 22 | 5 | 27 |
| Methanol | 62 | 49 | 8 | 57 |
| Hydrogen | | | | |
| MH | | 8.2 | 764 | 772 |
| HP(15MPa) | 670 | 8.2 | 755 | 763 |
| LH$_2$ | 115 | 8.2 | 65 | 73 |
| Battery(※) | | | | 1360 |

(※) Assuming that the energy density is 40 Wh/kg and the conversion efficiency of the battery to the power is 5 times higher than gasoline engine.

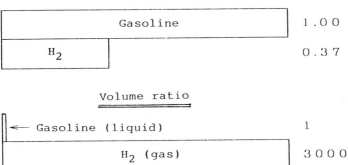

Fig 5.2.2 Transportability of hydrogen.

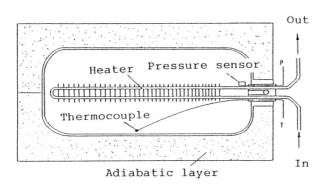

Fig 5.2.3 High pressure $H_2$ tank with temperature control unit.

temperature inside of the tank increases to be higher than 100 K, corresponding pressure becomes too high and danger arises.

Density of $LH_2$ is about one-tenth of gasoline density. Although Calorific value of hydrogen is 120,000 KJ/kg, which is about 2.7 times as large as the gasoline value, Calorific value per 1 $\ell$ of $LH_2$ is only 0.26 of that for gasoline. This means that under the same amount of energy generation, $LH_2$ requires only 1/2.7 in mass but nearly 4 times in volume as much as gasoline. In addition, since boiling point of $LH_2$ is 20 K, thermal insulation is very critical. Usually, $LH_2$ tank has double layer with vacuum space and super-insulation device in it. In the case of large tanks which allow a wide space between two layers, evaporation loss from the tank is a couple of tenth % per day. However, small $LH_2$ tanks for automobiles, whose storage volume is 100 to 200 $\ell$, have narrow vacuum space because of compactness, thus the evaporation loss are larger, i.e. 3 % per day at the present technology level and 1 % per day as a future target. It should be also noted that because of small density, level meter tends to be less accurate at the present situation. Evaporated hydrogen during engine operation can be made use of as fuel by mixing it with the air at the intake port. However, during engine halt, evaporated hydrogen should be blown out through a safety valve. In case the evaporated hydrogen can not escape from the tank, the tank pressure becomes so high that the tank breaks down and all hydrogen in the tank flows out. In order to avoid such an accident, double safety is usually provided.

The gelatinous mixture of $LH_2$ and slush hydrogen is denoted here by $S\ell H_2$, where the slush hydrogen is powdered solid hydrogen. Properties of 50 % $S\ell H_2$, which has 50 % slush hydrogen in weight, are tabulated and compared with $LH_2$ properties in Table 5.2.4. It is shown that 50 % $S\ell H_2$ density is larger by 15 % than $LH_2$ and that $S\ell H_2$ has higher cooling capacity. Since the $S\ell H_2$ is planned to be applied to hypersonic planes, where fuel is to be also used as a coolant for its engine and body, the higher cooling capacity is an advantage of $S\ell H_2$.

### 5.2.3. Development of $LH_2$ Pump and $LH_2$ Tank

### 5.2.3.1 Pump [5.2.2]

As mentioned before, a direct fuel injection method is preferred for hydrogen engines because power output can be increased and backfire can be prevented. To realize the injection method, supply of high pressure hydrogen is crucial. It is not practical to compress gaseous hydrogen by a compressor, as in the case of MH storage. On the other hand, only a small pump is necessary, if hydrogen is compressed in its

Table 5.2.4 Characteristic of slush hydrogen.

| | Triple point solid $H_2$ | 50% sl $H_2$ | Triple point $LH_2$ | 1 atm $LH_2$ |
|---|---|---|---|---|
| Temperature (K) | 13.80 | 13.80 | 13.80 | 20.30 |
| Pressure (kPa) | 7.03 | 7.03 | 7.03 | 101.40 |
| Density (kg/m$^3$) | 86.67 | 81.54 | 77.06 | 70.81 |
| Incase in cooling capasity from $LH_2$ | 111.80 | 82.72 | 53.60 | 0 |

liquid state. For example, a single cylinder reciprocating pump with 15 mm piston diameter, 15 mm stroke and 1000 rpm speed is sufficient for 100 ps engine. Selection and composition of materials are critical for $LH_2$ pump.

Self-lubricating material shuld be selected for the rubbing surfaces of a piston and a cylinder because no lubricants can work at 20 K. If a metal pump is used for $LH_2$, it will be immediately out of order. However, it was found out that synthetic resins such as Tefron or Polyimide had their hardness increase at low temperature, and were highly resistive to friction and wear without lubricants as far as without large pressure was acting on the contact surface. Since these materials have much larger thermal expansion coefficients $\alpha$ than metal, their composition with metal parts is also important. Figure 5.2.4 shows one of the best compositions, where the cylinder is made of metal, and the piston surface and its inner part are made of synthetic resin and invar (extremely low $\alpha$), respectively. The piston gap could be kept to be 1.5 ~2 $\mu m$ regardless of temperature if this composition was adopted and thickness of the synthetic resin collar was determined according to Equation (1).

Among various problems that have been studied and resolved, the most difficult one was so called "Gas Compression", by which the pump abruptly became occupied by gaseous hydrogen and $LH_2$ ceased to deliver. It was finally revealed that slight inclination of the piston and resultant pressure force generated large friction heat and increased evaporation of $LH_2$ during the suction stroke of pump.

A rod connecting the pump piston and the crank shaft is desired to be long and slender because $LH_2$ can be easily taken into the pump from bottom part of the tank, and heat inflow through the rod can be minimized. However, if the piston is reciprocating as usual case, compression force acts on the rod and its diameter has a certain limit for avoiding deformation. Thus, the pump was modified in such a way that the cylinder was reciprocating instead of the piston, as shown in Figure 5.2.5. As a result of the modification, tensile force acted on the rod, and the rod was able to be made much more slender without any deformation.

5.2.3.2 Improvement of Tank Insulation

Figure 5.2.6 illustrates a $LH_2$ tank on board inside of a passenger car trunk room and its fuel supply system. The tank has oval cross-section, and its inner volume is 82 $\ell$, which is equivalent to 21 $\ell$ gasoline. The $LH_2$ pump is driven by a crank mechanism. The pressurized $LH_2$ becomes $GH_2$ immediately after flowing out of the pump, and the $GH_2$

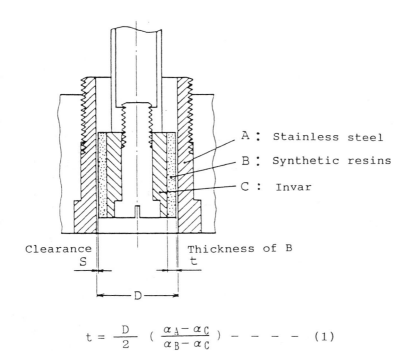

$$t = \frac{D}{2} \left( \frac{\alpha_A - \alpha_C}{\alpha_B - \alpha_C} \right) - - - - (1)$$

Fig 5.2.4  Material combination of piston and cylinder of piston.

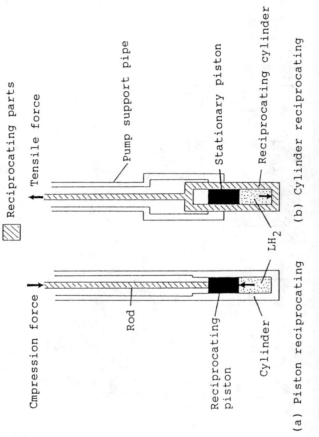

(a) Piston reciprocating    (b) Cylinder reciprocating

Fig 5.2.5  Cylinder reciprocating $LH_2$ pump.

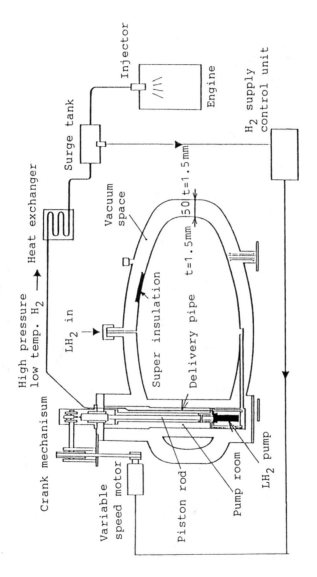

Fig 5.2.6 LH$_2$ tank with a pump.

is heated up to room temperature at a heat exchanger. Although injected hydrogen with higher temperature provides the engine with larger energy, this effect is not a large. In order to maintain pressure inside of the surge tank for various rate of injection amounts, i.e. various engine output powers, DC motor speed of the pump is adjusted by electronic controllers.

Since the upper part of the pump room of the tank had a contact with the atmosphere, heat inflow through the pump room amounted to 2.5 times as much as that through the other part of the tank. Until very recently, R & D efforts had been focused on improvements in performance and efficiency of the engine system including the pump. However, it was revealed that tank insulation was as important, and research work was initiated for improving the insulation effect.

Measurements of temperature distribution were made for the gas inside of the pump room and for the stainless steel wall. It was found out from the measurement that most of the heat inflow was transferred to $LH_2$ through the metal wall, but that convective heat transfer in the pump room could not be neglected. As the first countermeasure against the heat inflow, diameter of the piston rod and that of the pump room wall were made smaller, and their thickness were also reduced. Entrance of the pump room should be exposed to the atmospher because the pump is installed to and removed from the tank through the entrance. In order to increase thermal resistance against the heat inflow, the pump room is usually designed to be a long neck, as shown in Figure 5.2.7 (a). However, such a design was not convenient for automobile use, the long neck was installed inside of the tank, as shown in Figure 5.2.7 (b). In this design, mechanism for $LH_2$ level-down was essential because without the mechanism, $LH_2$ level in the pump room would have raised to the $LH_2$ level in the tank and the effect of the long neck would have been diminished from $L_1$ to $L_2$. Various types of such mechanism were tested. Among them, the easiest and practical method was to push down the $LH_2$ level by means of gas pressure in the pump room.

Figure 5.2.8 illustrates the tank structure with further improvement. In this case, the pump room was positioned at one side of the tank and was contained inside of vacuum space. Delivery of $LH_2$ to the pump was made through a suction pipe shown in Figure 5.2.8. This tank arrangement allowed the heat flow path to become much narrower and longer. In addition, to suppress convective heat transfer inside of the pump room, Teflon fins were attached to outer surface of the support pipe for the pump.

Reduction of heat inflow after each countermeasure described above is

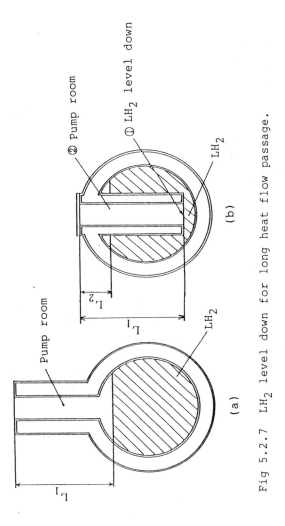

Fig 5.2.7 LH$_2$ level down for long heat flow passage.

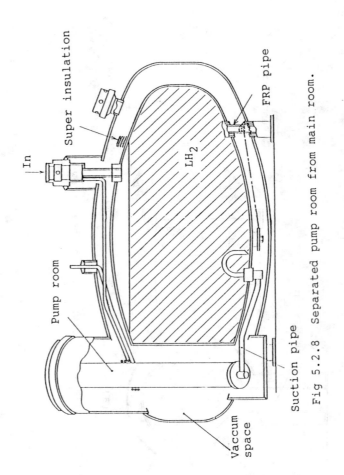

Fig 5.2.8  Separated pump room from main room.

tabulated in Table 5.2.5. After all of these modifications were implemented, the amount of heat inflow was decreased to only 8 % of that without them (initial stage). As a result, the ratio of the heat inflow through the pump room and that through the tank wall became 0.2, which was initially 2.5, and the evaporation loss from the pump room was reduced to 1 % per day.

### 5.2.4. Research and Development of Hydrogen Engine

#### 5.2.4.1 Premixture Combustion and Backfire

Hydrogen engine research activities are conducted all over the world, and in most cases, hydrogen fuel is mixed with air prior to entering the engine cylinder. This method is adopted for all automobile hydrogen engines except the one developed at Musashi I.T.. Although the pre-mixing method has several serious disadvantages as pointed out before, there seem following reasons for preferring this method to the injection method.
(i) Engine structure becomes simpler than that of gasoline engines, and a carburetor can be eliminated.
(ii) Extremely lean combustion is possible.
(iii) The injection method requires $LH_2$ pump, which may cause difficulties.
(iv) Use of $LH_2$ necessiates additional technologies.
(v) MH is considered as a primary candidate for storage.

At Musashi I.T. Engine Laboratory, premixture combustion hydrogen engines were developed for 7 years since 1970. In these engine systems, hydrogen was able to be supplied to the intake pipe only through a flow control valve without a carburetor. However, when hydrogen flow rate was increased and combustion gas temperature became higher, flames with light blue color flowed back through the intake valve, then the engine stopped with an explosive noise. THis is the backfire, which makes the normal engine operation impossible.

There are several potential causes of the backfire, but the primary cause has not been determined yet. The followings are two of the most likely causes.
(i) Residual combustion products and high temperature particles of carbonized oil are floating in the cylinder or attached to the combustion chamber wall, and they ignites $H_2$-air mixture as soon as the mixture comes in contact with them.
(ii) Hot spot such as pointed head of the spark plug electrode ignites $H_2$-air mixture.

According to experimental results at Musashi I.T. [5.2.3] it was found out that the spark plug pointed head melted down when engine operation with the backfire was continued. It was also revealed from the

Table 5.2.5  Decreasing of heat inflow through the $LH_2$ pump and pump room.

| Countermesure | Ratio of heat inflow |
|---|---|
| Initial stage. | 100 % |
| Smaller room and $LH_2$ level down to the bottom of the room. | 29 % |
| Fined support pipe to stop the vertical gas flow. | 22 % |
| Transfer the pump room from center to side of tank. | 8 % |

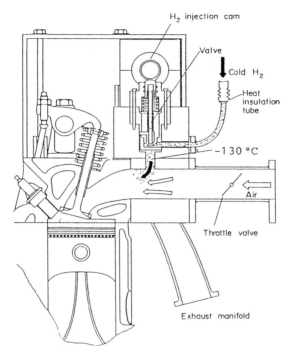

Fig 5.2.12 Low temperature $H_2$ supply system in case of MUSASHI-2 (1975).

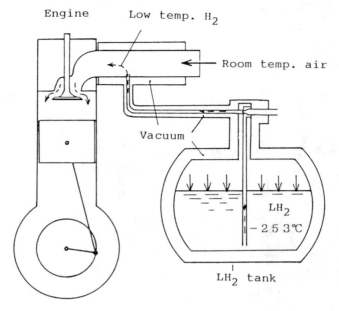

Fig 5.2.13  Intake port $LH_2$ supply system.

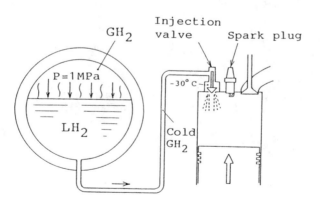

Fig 5.2.14  A low pressure $H_2$ injector system without $LH_2$ pump.

LH$_2$ pump was applied to a two-stroke engine, which is illustrated in Figure 5.2.15. As shown in this figure, the injection valve was driven by cam mechanism because of the long injection period. Application of hydrogen to a two stroke engine has following advantages.

(i) A two stroke engine has neither an intake valve nor an exhaust valve, thus its structure is more simple than that of a four cycle engine. Since a two stroke engine accompanies combustion at each revolution, higher power output can be expected.

(ii) In the case of a gasoline two stroke engine, part of fuel-air mixture passes through to an exhaust port without combustion, which results in large loss and increases HC in the exhaust gas. If gasoline, is injected into the cylinder during compression stroke (after exhaust port closed) ignition can not be stable. On the other hand, hydrogen-air mixture can be ignited throughout very wide range of mixing ratio. Therefor, in the case of a hydrogen two stroke engine, it becomes possible to inject hydrogen after an exhaust port is closed, to ignite the mixture by spark and to burn it completely. As a result, no "pass-through" of the fuel is expected, which should be a remarkable advantage.

Figure 5.2.16 shows comparison of measured thermal efficiencies between a hydrogen two stroke engine and a gasoline counterpart. The higher efficiency for the hydrogen engine is mainly because of no pass-through. However, in this low pressure injection engine, preignition occurred and normal engine operation was hindered at about the same output power as the maximum output of the gasoline engine. Relatively simple countermeasure, which is to inject cold hydrogen at $0 \sim -50°C$, was able to prevent the preignition in case of LH$_2$ fuel. As a result, approximately 20 % higher output power than the gasoline engine was obtained. In addition, as shown in Figure 5.2.17, reduction of exhausted NOx was noticeably large. It should be noted, however, that if injected hydrogen temperature was lower than $-100$ °C, lubrication oil in the injection system might freeze. It was Musashi 3 in 1978 that was equipped with a 550 cc mini-car two-stroke hydrogen engine, as shown in Figure 5.2.18.

### 5.2.4.4 High Pressure Hydrogen Injection [5.2.5]

Major objectives for adopting the high pressure injection for the hydrogen engine are as follows:
(i) To completely prevent abnormal combustion such as backfire, preignition and knocking.
(ii) To apply hydrogen fuel to large engines.
(iii) To increase compression ratio so that increasing a engine performance.

In the case of the high pressure injection, hydrogen injection starts

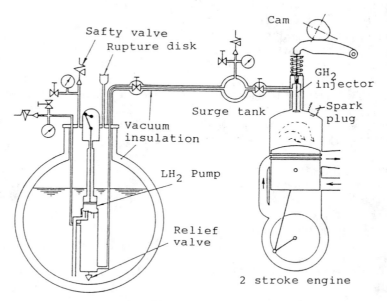

Fig 5.2.15  $H_2$ fuel system in MUSASHI-3 and 4.

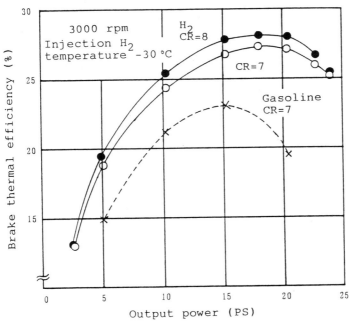

Fig 5.2.16 Performance of $H_2$ 2-stroke 550 cc engines.

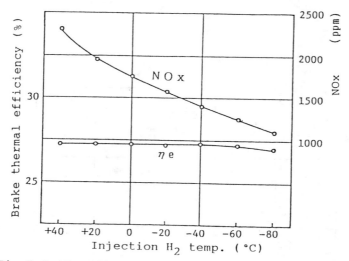

Fig 5.2.17 Effect of injection-$H_2$ temperature on NOx emission.

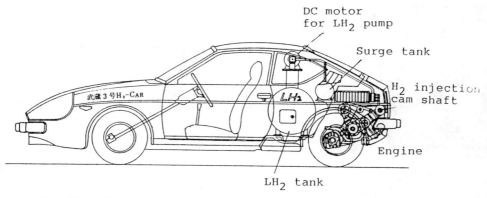

Fig 5.2.18 MUSASHI-3 (Mini car with 2-stroke, low press. $H_2$ injection engine.)

near the end of compression stroke and lasts for relatively short period, which is the same as conventional diesel engines. However, the following modifications were necessary to realize the high pressure injection method.

Injenction Mechanism

Since hydrogen should be injected during combustion, the injection pressure has to be higher than the maximum engine pressure, and the injection period has to be short. The injection pressure was raised to 8 MPa ($\simeq$ 80 kg/cm$^2$). The modified injection mechanism is illustrated in Figure 5.2.19. The pump ⓟ of an original diesel engine supplied high pressure oil to the nozzle ①. This high pressure force was transferred from the nozzle via the rod ② to the hydrogen valve ③, which was, as a result, pushed down and opened, then injecting the hydrogen through the hole ④.

Ignition Device

Since self ignition temperature of hydrogen is 580 ℃, which is too high for compression ignition, an ignition device is inevitable. An electically heated platinum wire was first tried, but despite its high melting point, its poor durability resulted in burn-out after a short period of use at higher temperature than 900 ℃. It was revealed that the burn-out was because the platinum wire was softened and lengthened. On the other hand, as shown in Figure 5.2.20, the ignition hot surface was preferred to be higher than 900 ℃ from the viewpoint of engine performance. Therefore, a more durable ceramic glow-plug was used as the hot surface. Figure 5.2.21 illustrates relative positions of the combustion chamber, the injection valve and the ignition hot surface.

Compression Ratio

In the case of hot surface ignition, compression ratio need not be raised even at cold start. Since the optimal compression ratio for 8 MPa injection pressure was 12 : 1 $\sim$ 15 : 1, as indicated in Figure 5.2.22, the ratio 12 : 1 was chosen for the hydrogen engine. Figure 5.2.23 a photograph of the medium size truck with the high pressure injection type hydrogen engine which has the feature described above.

5.2.5. Recent Research at Musashi I.T.

5.2.5.1 Visualization of Hydrogen Combustion [5.2.6]

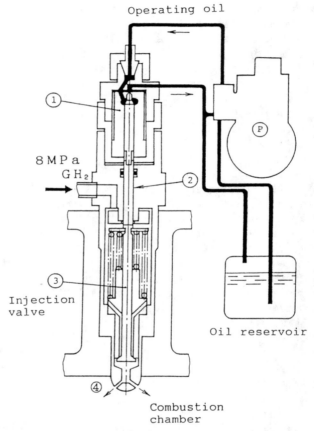

Fig 5.2.19 High pressure H$_2$ injection apparatus.

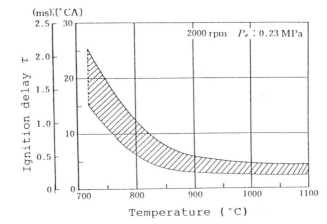

Fig 5.2.20 Effect of hot surface temperature on the ignition delay $\tau$.

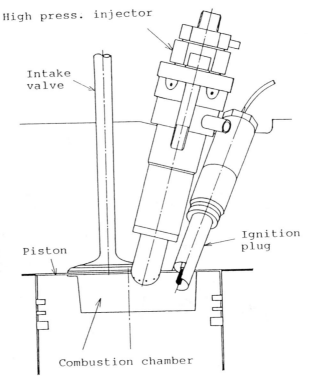

Fig 5.2.21 Combustion chamber of MUSASHI-6 and 7.

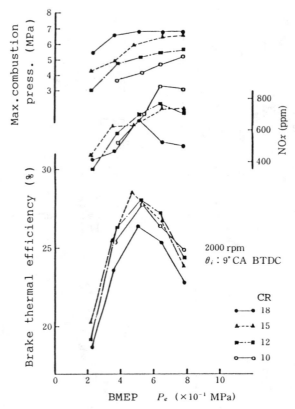

Fig 5.2.22 Effect of CR on the engine performance.

Fig 5.2.23 Truck with high pressure $H_2$ injection engine.

It is important to grasp the way of flame propagation for making better understanding of the hydrogen combustion. However, it is also difficult to visualize the hydrogen combustion because its flame is colorless and it has little ionization. Recently, the hydrogen combustion inside a constant volume combustor instead of a reciprocating engine was visualized by high speed schlieren pictures. Figure 5.2.24 is the typical sets of pictures, where hydrogen was injected from the port ① toward the glow plug ②. The injection started at 0.26 ms from the volve open (t=0), and one of the injected jet was burning ( blackened in the picture ) at 0.78 ms. This flame, however, did not propagate to the neighboring jet until 1.69 ms, and after that, the flame propagation took place rapidly. These two moments corresponded to the points A ($\tau_A$ = 0.78 ms) and B ($\tau_B$ = 1.69 ms) of the pressure indicator diagram in Figure 5.2.25. In such a two-stage combustion, the glow temperature altered length of $\tau_A$ but did not affect length of $\tau_B$. Since rate of the following pressure rise and strength of the pressure vibration were related to the $\tau_B$ value, change of the glow temperature did not help their reduction.

Ignition by electric spark has advantages such as very small ignition energy and very wide range of mixing ratio at which the mixture can be ignited. Visualization of its ignition is shown in Figure 5.2.26. The tip of the hydrogen jet was most easily ignited as (a). As the ignition timing was delayed, the ignition became more unstable as (b), and the bottom part of the jet could not be ignited as (c). It was also found out from this visualization that smaller distance between the injection port and the spark gap ( smaller X in the figure ) enabled the jet tip to reach the gap earlier, made the ignition delay shorter, and realized more gradual pressure rise. This result indicates that in the case of applying the spark ignition to the hydrogen injection type engine, precise positioning and carefully adjusted timing will be necessary.

5.2.5.2 Decrease in Number of Mols after Hydrogen Combustion [5.2.7]

In the case of petroleum fuel, number of mols increases after combustion ; for example $C_8H_{16}$ combustion results in 5 % increase as shown in Figure 5.2.27. On the other hand, after hydrogen combustion, number of mols decreases by 15 %. This result often leads to misunderstanding that hydrogen combustion engine efficiency is inevitably low. However, since molar specific heat does not vary with gas species, temperature of combustion products should increase if number of mols decreases. It seems that such temperature increase compensates for pressure drop caused by the mole number decrease. Figure 5.2.28 shown the calculated maximum pressure for various fuels under constant volume combustion with temperature dependence of molar

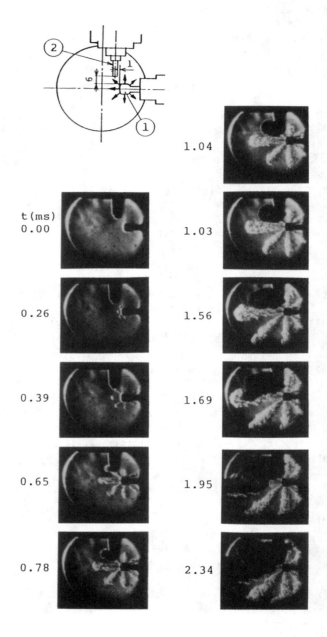

Fig 5.2.24 Ignition and flame propagation of $H_2$ jet. (Schlieren method)

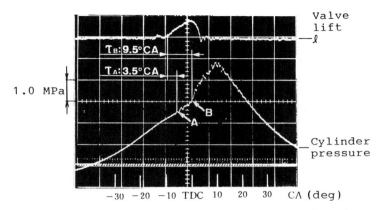

Fig 5.2.25 Cylinder pressure of tow stage combustion.

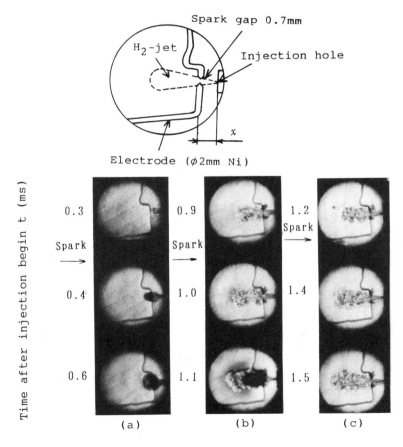

Fig 5.2.26 Inflammability of $H_2$ jet with spark.

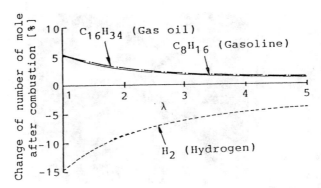

Fig 5.2.27  Change of mol number after combustion.

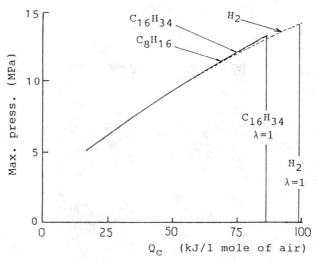

Fig 5.2.28  Effect of mol number change on the explosion pressure.

specific heat taken into account. As clearly demonstrated in this figure, decrease in number of mols does not affect hydrogen engine performance.

Furthermore, injected hydrogen amounts to 42 % of air in terms of mol number under the stoichiometric ( $\lambda=1$ ) condition. Pressure increase and resultant increase in output power attributing to the hydrogen injection turn out to be 3.5 % and can not be overlooked.

### 5.2.5.3 Improvement in Mixture Formation [5.2.7]

It was suggested that mixture formation after hydrogen injection be improved for higher engine performance. Difficulties in the mixture formation lie in the following characteristics of hydrogen. First, since hydrogen density is only 1/14.5 of air density, injected hydrogen jet at about 100 atms, which is optimum with respect to the pump design, can not sufficiently penetrate into the air. Secondly, although hydrogen has high diffusivity, it is not high enough to promote mixture formation.

The instantaneous heat generation rate $dQ/d\theta$ at each crank angle $\theta$ and the integrated heat generation $q_e$ up to the crank angle $\theta$ were calculated from the measured pressure in the cylinder. Their results are shown in Figure 5.2.29, where it was indicated that higher injection pressure improved combustion performance. In addition, use of a shroud valve as an intake valve augmented swirl flow and resulted in higher combustion performance, as shown in Figure 5.2.30.

### 5.2.6. Safety of Hydrogen

It seems that dangerous aspects of hydrogen is exaggerated. As a result, various regulations against hydrogen are causing unnecessary difficulties to its research and developments. Thus, more reseach activities on hydrogen safety and its countermeasures will be desired to accumulate more scientific data and to establish appropriate supervision and regulations.

From the author's 20 year experience of hydrogen research, it may be concluded that hydrogen is in general as dangerous as gasoline or propane. In fact, among a number of current and previous research activities on hydrogen engines and hydrogen fueled vehicles, there have been no big accident reports. However, it should be noted that each fuel has different characteristics with respect to its safety.

From the following viewpoints, hydrogen is safer than other fuels.
(i) Self ignition temperature of hydrogen is high, therefore

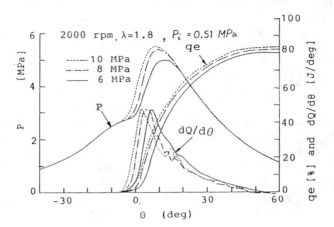

Fig 5.2.29 Effect of injection press. on the combustibility.

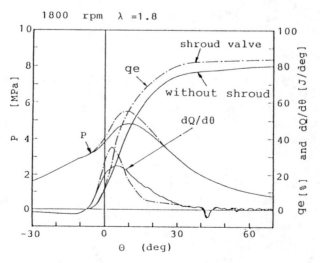

Fig 5.2.30 Effect of shroud valve on the combustion.

troposphere is the stratosphere until 50 km and above the stratosphere are the memosphere, the thermosphere and the magnetosphere (exosphere). The structure of the troposphere and the stratosphere are shown in Fig. 6.1.2. As seen from the figure, the absolute amount has the extreme at an altitude of about 25 km and the volumetric ratio has the extreme at an altitude of about 30~40 km. Generally the discharged gas molecules will be convected and diffused to the stratosphere through the troposphere for a period of one to several years in the case where there is no decomposition and absorption in the troposphere. The mechanism of the production of the stratospheric ozone layer has been explained elementarily by the following four reactions (Chapman mechanism). [6.1.2]

$O_2 + h\nu \rightarrow 2O$ (1)  $O + O_2 + M \rightarrow O_3 + M$ (2)

$O_3 + h\nu \rightarrow O + O_2$ (3)  $O + O_3 \rightarrow 2O_2$ (4)

Concerning the decomposition cycle of the ozone by the effect of the trace gases, the amount of the trace gas X is included in the following reactions.

$X + O_3 \rightarrow XO + O_2$ (5)  $XO + O \rightarrow X + O_2$ (6)

[ The addition of the reaction (5) and (6) is equivalent to the reaction (4).]
The trace gas X, OH and H, NO and Cl make the major contributions, which constitute the HOx cycle, NOx cycle and ClOx cycle, respectively. The main sources of OHx and NOx in the stratosphere are $H_2O$ and $N_2O$ respectively, which would be discharged to the atmosphere by natural phenomena. The difference of the solar radiation spectrum in the stratosphere and the troposphere is the strength of the ultraviolet rays of short wavelength (wavelength of 280~320nm), since in the stratosphere the ultraviolet rays of C region (wavelength below 280nm) were absorbed perfectly and the ultraviolet rays of B region (wavelength of 280~320nm) which would be also harmful to living things were absorbed partially. When the CFC entered the stratosphere where the strength of the ultraviolet rays was increased in the shorter wavelength of 190~210nm, called the window in the ultraviolet region, CFC was decomposed by the absorption of the ultraviolet rays to discharge the atoms of chlorine. The stratospheric ozone problem is the decrease of the amount of ozone by decomposition through the cycle of ClOx caused by the discharged chlorine atoms described above and there is the possibility of increasing the strength of the ultraviolet rays of B region which reached the surface of the earth. This is caused by the decrease of the amount of ozone in the stratosphere. It was also clarified experimentally in the artificial conditions of the stratosphere that the ozone has been decomposed by the existence of CFC.[6.1.3] The total concentration of chlorine in the stratosphere is estimated to be larger than 3 ppb at the present stage and over three quarters would be discharged artificially.

Although the amount of the decrease of the ozone due to chlorine atoms generated by CFC would be difficult to estimate, the total amount of CFC discharged in the atmosphere would reach about 20 million tons at the present stage and additionally one million tons of CFC have been discharged annually. Since CFCs are estimated to be very stable and have a lifetime in the atmosphere of about one century, almost all of the discharged CFC have been

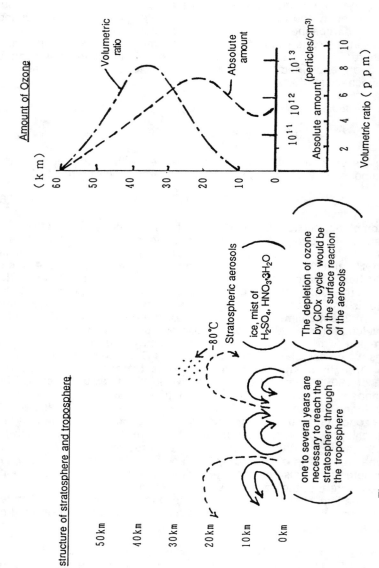

Fig.6.1.2. Structure of stratosphere and troposhere and distribution of ozone

estimated to be stored in the troposphere and only 10% of the total amount are estimated to have reached to the stratosphere.

Concerning the measure of influence on the amount of ozone, ODP (Ozone Depleting Potential: the relative amount of influence on the depletion of the ozone compared with that of CFC-11 for a unit weight) means the weight of material necessary to reduce 1% of the total amount of ozone and is calculated according to the following model. Although calculating ODP would be equivalent physically to the estimation of the decomposition rate in the atmosphere, the computer simulation has been carried out by changing the parameters and the influencing factors such as the effect of convection, the distribution of the concentration of reactive materials, the temperature distribution of the atmosphere and seasonal fluctuations. The most simple model is a one-dimensional model, where the atmosphere is estimated to be uniform in the horizontal direction and distributed only in the vertical direction. By dividing the altitude up to 60 km into over thirty subregions and assuming uniform distribution in the subregions, the conservation equations including the transport phenomena and chemical reactions are to be solved among the subregions. The boundary conditions on the surface would be the discharge rate of the materials at the present, and at the top of the atmosphere the equilibrium of the photochemical reactions would be assumed. The distribution of the eddy diffusivity has been assumed in order to match the distribution of $CH_4$ or $N_2O$ with the measured distributions. As the chemical species, about forty species such as $O_3$, O, NO, OH, Cl, various kinds of CFC, H, ClO, etc. have been considered. The temperature distribution assumed the measured distribution for the most simple case and for the more complex models it was calculated by solving the energy equations containing the absorption of ultraviolet radiations by ozone and the absorption and the emission of infrared radiations. The intensity of the solar radiation was given for the specified latitude and longitude. The photo-dissociation constants have been determined depending on the altitude, the temperature and the density. The averaged conditions of clouds have been used depending on the altitude. Then, by solving the conservation equations among the subregions, the steady state distributions of chemical species and the intensity of the solar radiation have been obtained. The total amount of the ozone in the column of the atmosphere, as shown in Fig. 6.1.3, was calculated according to the steady state distribution. As explained above, the decrease of the total amount of the ozone was then calculated and compared with that for CFC-11 to obtain the value of ODP. As seen from the figure, the solutions have some possibility of being influenced by the amount of usage and social and economical conditions, since the boundary conditions were assumed to be the present rate of production. Furthermore, since the steady state solutions have been obtained to realize the saturated concentration, the necessary period for obtaining the steady state conditions were mostly 50~100 years. In other words, for the case of CFC, the lifetime would be the order of 100 years. There have been more complex models of two-dimensional models and three-dimensional models. In the two-dimensional models, the regions were divided in the direction of the latitude and in the three-dimensional models the surface of the earth as divided into many meshes of subregions, but the meshes were still rough for the case of Japan, to be divided into several subregions.

The values of ODP obtained by the above procedures are shown in Table 6.1.1.[6.1.12] As understood from the above explanation, a significant digit

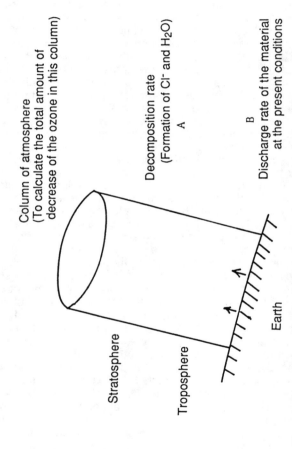

Fig.6.1.3. Boundary conditions of numerical simulation of ODP

Table 6.1.1. ODP and GWP of CFC, HCFC, HFC, etc

| Species | ODPs | | GWPs |
|---|---|---|---|
| | | Montreal Protocol | |
| CFC-11 | 1.0 | 1.0 | 1.0 |
| CFC-12 | 0.9-1.0 | 1.0 | 2.8-3.4 |
| CFC-113 | 0.8-0.9 | 0.8 | 1.3-1.4 |
| CFC-114 | 0.6-0.8 | 1.0 | 3.7-4.1 |
| CFC-115 | 0.3-0.5 | 0.6 | 7.4-7.6 |
| HCFC-22 | 0.04-0.06 | | 0.32-0.37 |
| HCFC-123 | 0.013-0.022 | | 0.017-0.020 |
| HCFC-124 | 0.016-0.024 | | 0.092-0.10 |
| HFC-125 | 0 | | 0.51-0.65 |
| HFC-134a | 0 | | 0.24-0.29 |
| HCFC-141b | 0.07-0.11 | | 0.084-0.097 |
| HCFC-142b | 0.05-0.06 | | 0.34-0.39 |
| HFC-143a | 0 | | 0.72-0.76 |
| HFC-152a | 0 | | 0.026-0.033 |
| $CCl_4$ | 1.0-1.2 | | 0.34-0.35 |
| $CH_3CCl_3$ | 0.10-0.16 | | 0.022-0.026 |
| halon 1301* | 7.8-13.2 | 10.0 | |
| halon 1211* | 2.2-3.0 | 3.0 | |
| halon 2402* | 5.0-6.2 | to be determined | |

would be one digit due to the difference of the values depending on the models. The calculation procedure and the values of GWP will be discussed in the next subsection.

### 6.1.2 Global warming effects of CFC

The global warming effects of trace gases in the atmosphere are also important in addition to the global warming effects of $CO_2$. This is because the infrared absorption values of CFC are large especially in the region of the wavelength of $8{\sim}12\mu m$ which is called the window region of the atmosphere. Physically, these absorption characteristics refer to the infrared absorption by the vibration energy of the bond among F-C-F and Cl-C-Cl. Therefore, CFC gases are transparent to the solar radiation of which a stronger wavelength is in the visible region and which is equivalent to the radiation of 5750K, but CFC gases would absorb the radiation from the surface of the earth of which a wavelength is in the infrared region and which is equivalent to the thermal radiation of 300K or CFC gases would change the concentration of the other gases which have some effects on the global warming. In the actual atmosphere there are many global warming gases such as steam, carbon dioxide, ozone, methane and $N_2O$ and also cloud and aerosols have some effect on the global warming. Therefore, the surface temperature of the earth, which is estimated to be about -20°C without the existence of the atmosphere, has become about 15°C with the global warming effects.

Concerning the relative measure of the global warming effects, GWP (Global Warming Potential: relative global warming effects for a unit weight compared with the effects of CFC-11) has been calculated for many kinds of gases by the following methods. GHP (Greenhouse Potential) is the nearly same measure compared with the effects of CFC-12. The calculation procedure was basically the same as that for calculating the values of ODP. The temperature distribution and the temperature of the surface were calculated by considering the solar radiation, radiative heat transfer, albedo of the earth (reflectivity of the surface of the earth), convective heat transfer between the surface and the atmosphere, eddy diffusivity and the change of the amount of the steam and clouds. The most simple model would be the one-dimension energy balance model. In this model, the temperature distribution in the troposphere and the surface temperature was assumed to be constant and the temperature distribution in the stratosphere was calculated for the photochemical equilibrium conditions of the chemical species. This temperature distribution would be modified by the change of the concentration of the species due to the existence of CFC. Then, the radiative heat flux to the surface of the earth was calculated and the change of the heat flux was divided by the climate feedback parameter which was in the range of $1{\sim}4$ $W/m^2K$ and the change of the surface temperature was calculated. By this change of the surface temperature the value of the GWP was calculated. There were also more complicated two-dimensional and three-dimensional models. One of the most important boundary conditions would be the change of the surface boundary conditions for the increase of the surface temperature. The surface boundary conditions include the change of the albedo of the surface, the change of the generation of steam and the change of the generation amount and the shape of the clouds for the change of the surface temperature. Another important boundary condition would be the magnitude of the absorption by the sea which would have a large amount of heat capacity. However, to estimate the above boundary conditions would be very difficult and even the reliability and the magnitude of the error was difficult to be estimated for the numerical simulations as shown in Fig. 6.1.4. Furthermore, the boundary conditions could not yet contain the effect of the mountains and the surface roughness. GWP was the steady state value for the boundary conditions of the present production rate, similar to the case of ODP. The material that had a smaller decomposition rate had a larger amount of the steady state concentration and thus had a larger value of GWP and a longer lifetime. The calculation of the unsteady equations showed the change of the surface temperature for the concentration change of CFC from 0 ppv to 1 ppv, such as 0.17K for CFC-12, 0.14K for CFC-11 and 0.05K for HCFC-22, which enabled the transient change of the global warming effects of CFC.

As explained above, it is still very difficult to estimate the reliability of the numerical simulations and the magnitude of the error for both the calculations of ODP and GWP. Also there are many uncertainties for the boundary conditions. However, by considering the fact that the concentration of CFC in the atmosphere has been increasing and that the decomposition of CFC has been measured in the stratosphere, to estimate the future trends of the atmosphere accurately as much as possible and to carry out a practical policy for the global environmental protection is important. As was carried out in the IPCC (Intergovernmental Panel on Climate Change), scientific assessments on the various kinds of independent research and calculations have been carried out to clarify the common conclusions for the different equations and for the different boundary conditions of the independent numerical simulations, though they have had many problems and uncertainties. These scientific assessments are based on the opinion that, even though there would be many problems in the reliability and in the uncertainties, the estimation of the future trends of the environment should be made by collecting, reviewing and assessing all results of scientific research conducted so far to make clear the commonly derived conclusions for the different equations and for the different boundary conditions for taking care of the global environmental protection in advance. According to the above opinion, the numerical simulations of estimating the future trends with some uncertainties have become important, especially for assisting the determination of future policies, which would be a completely different role of the research and the numerical simulations.

However, the revision of the numerical simulations to remove the uncertainty is very important and the experimental research and the actual measurements are also important and should be promoted.

### 6.1.3 Regulations of CFC and alternatives

Concerning the environmental influences of CFC and the regulations of CFC, there were many international meetings, details of which are explained below. The Montreal Protocol on substances that deplete the ozone layer became effective from January 1989 and therefore 1989 has been called "The first year of CFC regulations". According to the Montreal Protocol, for the five specified CFCs (CFC-11,12,113,114,115) the production and the consumption rates were frozen to the level of 1986 from July 1989, they were to be reduced to 80% of the level of 1986 by July 1993 and they were to be reduced to 50% by July 1998. For the halons of 1211, 1301, 2402, which are frequently used for

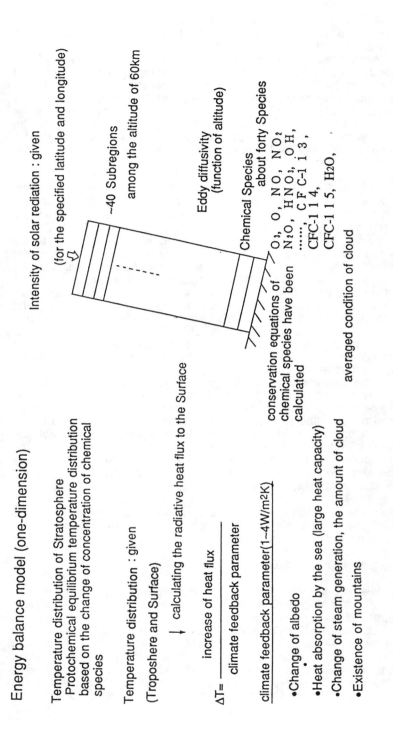

Fig.6.1.4. Characteristics of numerical simulation model of GWP (one-dimensional model)

fire extinguishers, the production rate was to be below 110% and the consumption rate was to be below 100% of the level of 1986 by 1991. The Montreal Protocol was ratified by 33 countries, e.g., U.S.A., U.S.S.R., Japan, etc. These countries were not permitted to import products containing CFC from the countries not yet ratified. In UNEP (United Nations Environmental Programme) the production of equipment containing CFC in overseas factories and the exceptional advantages for the developing countries have been discussed. But, after the start of Montreal Protocol many countries have insisted on the promotion and the strengthening of CFC regulations and in the Second Meeting of the Parties to the Montreal Protocol on Substances that Deplete the Ozone Layer, held in London July 1990, the phaseout of five specified CFCs and three specified Halons by the end of this century was decided. There are still some arguments of insisting on the uncertainty of the simulations for CFC regulations, and of the exceptional advantages for the developing countries.

Concerning the regulations of CFC, when Prof. Roland first presented the paper in 1974, the presentation of the law to the Congress regulating the usage of CFC had immediately been prepared in 1975. This was a similar situation as the first law regulating the emission of carbon dioxide which had already been presented to the Congress in 1989. Since the problems of NOx emission from the SST (Supersonic Transport) and "space shuttle" and the destruction of the ozone layer in the stratosphere have been investigated by CIAP (Climate Impact Assessment Program) in 1972~1974, several investigations were started immediately in some institutes such as National Science Foundation. In 1978 the usage of CFC for aerosol propellants was prohibited. Also, in Sweden, Canada, Norway and Thailand the usage of CFC for aerosol propellants has been prohibited one after another. In Japan, only an expression of making efforts to reduce the usage of CFC for aerosol propellants was conducted in 1980 and the main research topics of environmental pollution had focused on photochemical smog and acid rain at that time.

The necessary characteristics of the alternatives of the specified CFC are presently the following four items. (1). small value of ODP (2). small value of GWP (3). low toxicity (4).low flammability Also, low cost would be an important factor. Furthermore, the additional characteristics necessary for the refrigerants of the thermal cycles are the following: thermal stability, chemical stability, the characteristic of being corrosion-free for materials such as metals, high thermal efficiency of the ideal cycle on the diagram and the dynamic stability under the actual conditions with the periodical change of temperature and the lubrication oil. The following thermophysical properties are necessary for estimating the performance of the thermal cycle, and they have been measured and investigated in the Japanese national energy conservation project "Super Heat Pump Energy Accumulation Systems". They are critical temperature, critical pressure, critical density, vapor pressure, latent heat, density of gas and liquid, specific heat, thermal conductivity and viscosity.

The present amount of the usage of CFC reached about one million tons for the world (1985), and the amount for individual use was about 25% each for refrigerants, foaming agents, cleaning agents and aerosol propellants. In Japan the total amount has been about 15% of the world and the amount for the individual use was about 50% for cleaning agents, 25% for foaming agents, less than 20% for refrigerants and less than 10% for aerosol propellants. The presently promising environmentally acceptable fluorocarbons as alternatives of CFC are HCFC-123 as an alternative of CFC-11 and HFC-134a as an alternative of CFC-12. They are estimated to have lower values of ODP and GWP as shown in Fig. 6.1.1 and are located at the righthand side upper region without any hatching in Fig. 6.2.2 of the next section, which means nonflammable and non-toxic. The alternatives for the other CFCs are discussed in detail in the next section, but there would be the decomposition process for HCFC and HFC in the troposphere, which would contain the reaction of hydrogen bond with OH bond to produce the ionic decomposition with $H_2O$ and which would have the water soluble products of HCl, HF and HCOOH returned back to the surface with rain. HCFC-123 and HFC-134a would belong to the second generation or the third generation of fluorocarbons, since they contained the hydrogen atoms and have small values of ODP and GWP. In particular, the thermophysical properties for these promising environmentally acceptable fluorocarbons of HCFC-123 and HFC-134a have been measured comprehensively and published as a world-first data book.[6.1.13] Concerning the toxicity, an acute toxicity study using mice, rats and various kinds of bacteria, a subacute toxicity study, hereditary toxicity study, carcinogenicity test, reproduction toxicity test and a metabolic abnormality test have been conducted and these two alternatives have been reported to have no toxicity or rather small toxicity. Furthermore, the safety check for many kinds of HCFC and HFC has been conducted by the organization of PAFT (Program for Alternative Fluorocarbon Toxicity Testing).

On the occasion of the revision of the Montreal Protocol, some countries including the U.S.A. and Japan have expressed the abolition of the second generation of fluorocarbons which have larger values of ODP and GWP, by the year 2050. Since the bond between carbon and fluorine originally has strong infrared absorption characteristics, the fluorocarbons which would be easy to be composed have a smaller lifetime compared with the necessary diffusion time of 1~2 years to the stratosphere through the troposphere, or the fluorocarbons which would be soluble with water to be easily circulated by the function of the rain, will be researched and developed as the third generation of the fluorocarbons. Concerning the third generation of the fluorocarbons, some materials such as perfluorinated oxetane (cyclo-$CF_2$-$CF_2CF_2O$-) and perfluorinated dimethoxymethane ($CF_3$-O-$CF_2$-O-$CF_3$) have already been proposed[6.1.14], but there would still be many important and basic problems for obtaining the values of ODP and GWP of fluorocarbons as explained in the previous subsections. Therefore, further discussions and the accumulation of the data would be necessary before a detailed discussion of the third generation fluorocarbons can be submitted.

6.1.4 Alternative technologies and recovery technologies for the refrigerants of CFC

In this subsection, CFC and the second generation of fluorocarbons as the refrigerants of thermal cycles have been examined to clarify the measures for the improvement of the thermal cycle, recovery technologies of the refrigerants and alternative technologies. The destruction technologies of CFCs are explained in section 6.3.

One of the improvement technologies is to reduce the necessary amount of the refrigerants and the research subjects for this improvement are the reduction methods of leakage, perfect sealing with hermetic seal, estimation of the necessary and minimum amount of the refrigerants, the optimum cycle for the minimum amount of the refrigerants and the recovery technologies of the refrigerants for air conditioners, car coolers, refrigerators and large scale heat pump systems. Furthermore, especially for the second generation fluorocarbons which include no chlorine atom in its chemical formula, the selection of the lubricant oil and the solubility with the lubricant oils have been important problems not to deteriorate the performance of the compressor and the heat exchangers.

Concerning the recovery technologies of CFC, there are several kinds of recovery methods such as condensation methods and absorption methods. For the larger concentration of CFC, cooling the gas for making the condensate of CFC would be effective to treat the large amount of the gas. For the lower concentration of CFC, the absorption by use of the activated carbon and the absorption by use of special oil would be effective. The absorption method by use of the activated carbon has been effective even for the lower concentration of several hundreds ppm. Several equipments have been already developed and much efforts have been focused on decreasing the cost.

Concerning the alternative technologies, the change of the refrigerants from the fluorocarbons would be important. Water is one of the promising refrigerants and the expansion of the application temperature range of water would be an important problem. The improvement of the absorption heat pumps and the introduction of the chemical heat pumps are also important. Although the research and development for the improvement of the absorption heat pump have been active in the U.S.A. and Europe, the barrier would be high for the spread of the absorption heat pump in Japan, since the regulations for the use of ammonia have been severe. As for the chemical heat pumps utilizing the hydrogen, steam and organic refrigerants, the details are explained in section 10.4.

As an advanced refrigeration and heat pump cycle, Vuilleumier cycle utilizing the refrigerant of Helium will be explained below. This recently developed cycle was composed of a Stirling engine and a Stirling cycle refrigerator utilizing Helium. In one machine the Stirling engine was operated between the high temperature heat source and the thermal output temperature, and the Stirling cycle heat pump was operated between the thermal output temperature and the low temperature heat source. The structure composed of two displacer-type Stirling engines and the temperature-entropy diagram are shown in Fig.6.1.5. Since the thermal output would be the sum of the thermal output of the Stirling heat pump and the exhaust gas of the Stirling engine, the high performance will be expected. The COP for the heating case has been about 1.5 and the COP for the cooling case has been about 0.6~0.8 for the thermal output of several kW class at the present condition. But, the improvement of doubling the COP would be expected by decreasing the heat conduction loss.

REFERENCES
[6.1.1] M. J. Molina & F. S. Rowland, Nature, 249,810-2,1974.

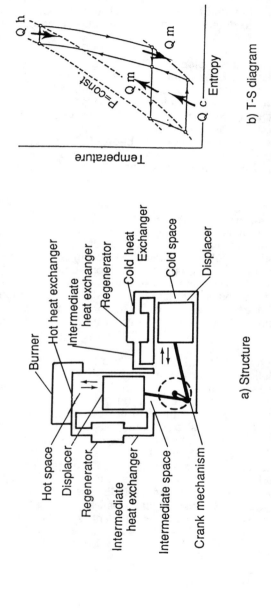

Fig.6.1.5. Structure and T-S diagram of Vuilleumier cycle

[6.1.2] S. Chapman, Mem. Roy. Meteorol. Soc., 3.103-25, 1930.
[6.1.3] H. Bando, Destruction of the stratospheric ozone, Thermal engineering in the advanced technology, 1989, p.76.
[6.1.4] Atmospheric Ozone (1985) WMO, p.1100.
[6.1.5] D. Wuebbles, J. Geophysical Research, 88-C2, 1983, p.1433.
[6.1.6] Protection of the ozone layer,NHK Books, 1989.
[6.1.7] N. Ototake, Research on the alternative fluorocarbons, kogyocyosakai, 1989.
[6.1.8] Refrigeration and air conditioning, 345,1990-1.
[6.1.9] T. Tominaga, Photo-chemistry of fluorocarbon, New fluorine chemistry, kagakusousetsu, 27, 1980, p.231.
[6.1.10] T. Suganami, et. al., Vuilleumier cycle heat pump, Proc.3rd IEA Heat Pump Conf., Pergamon Press, 1990, p.585.
[6.1.11] M. O. Mclinden & D. A. Didion, Quest for alternatives, ASHRAE J., Dec., 1987.
[6.1.12] Data submitted to the Nairobi meeting for the revision of Montreal Protocol, UNEP , 1989.
[6.1.13] Thermophysical properties of environmentally acceptable fluorocarbons---Hfc-134a and HCFC-123---, Japanese Association of Refrigeration, 1990.
[6.1.14] Progress on CFC substitutes discussed, C&EN, Sep. 3, 1990, p.8
[6.1.15] T. Shimazaki, Stratospheric ozone, Univ. Tokyo Press, 1989.
[6.1.16] K. Watanabe, Current status of thermophysical properties research on CFC alternatives, Proc. 3rd IEA Heat Pump Conf., Pergamon Press, 1990, p.263.

## 6.2 CFC ALTERNATIVES

**Takaji Akiya**
National Chemical Laboratory for Industry
Chemical Systems Division, Physico-chemical Properties Section
Agency of Industrial Science and Technology
Ministry of International Trade and Industry
Tsukuba Science City, Ibaraki 305, Japan

6.2.1. Introduction

The freon such as CFC-11 and CFC-12 have been broadly utilizing as a refrigerant, a foaming agent and a cleaning agent, because they are chemically stable, non-toxic and nonflammable. However, in view of the fact that the freon diffused up to the stratosphere destroys the ozone layer, the use of these materials have been restricted, and now research and development of a novel refrigerant which does not attack the ozone layer is expected. The necessary conditions of the novel refrigerant to be added to the common characteristics of the freons are as follows:
 (1) From view of "the ozone problem", the new refrigerant shall be free from chlorine atom. If the circumstance requires it, the refrigerant shall be made decomposable by introducing hydrogen atom thereinto.
 (2) The refrigerant shall be chemically and thermally stable.
 (3) The refrigerant shall be fire retardant, preferably nonflammable.
 (4) The refrigerant shall be non-toxicity.
 (5) It is important that the manufacturing process has already been established

As mentioned above, the alternative CFCs shall be so as to see in Figs. 6.2.1 and 6.2.2.

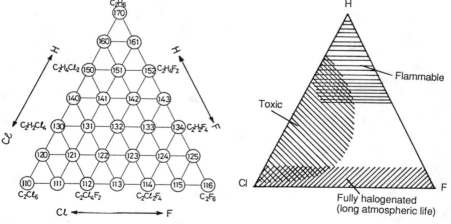

Fig. 6.2.1 CFC Derivertives of Ethane    Fig. 6.2.2 Characteristics of CFC Derivertives of Ethane

Fig. 6.2.1 indicates a state wherein hydrogens, fluorines and chlorines bond respectively to the substance comprising two carbon atoms which constitutes the CFCs nucleus. The general characteristic of these CFCs are indicated in Fig. 6.2.2 [6.2.1]. Considering the toxicity and flammability,etc., indicated in Fig. 6.2.2, the characteristics of the candidates being currently proposed as the alternative CFCs are shown in Table 6.2.1 [6.2.2].

Table 6.2.1 The ODP, GWP of CFC Althernatives

| Refrigerant | Molecular Formula | Boiling Point [°C] | ODP | GWP |
|---|---|---|---|---|
| HCFC-123 | $CHCl_2CF_3$ | 28.7 | 0.013~0.022 | 0.017~0.020 |
| HCFC-124 | $CHClFCF_3$ | -12 | 0.016~0.024 | 0.092~0.10 |
| HFC-125 | $CHF_2CF_3$ | -48.5 | 0 | 0.51~0.65 |
| HCFC-132b | $CH_2ClCClF_2$ | 46.8 | < 0.05 | < 0.1 |
| HFC-134a | $CH_2FCF_3$ | -26.5 | 0 | 0.24~0.29 |
| HCFC-141b | $CH_3CCl_2F$ | 32.0 | 0.07~0.11 | 0.084~0.097 |
| HCFC-142b | $CH_3CClF_2$ | -9.2 | 0.05~0.06 | 0.34~0.39 |
| HFC-152a | $CH_3CHF_2$ | -24.7 | 0 | 0.026~0.033 |
| HCFC-22 | $CHClF_2$ | -40.8 | 0.04~0.06 | 0.32~0.37 |

At present, mostly concerning the adaptabilities of the CFCs alternatives to the refrigerant of heat pump or refrigerator, the foaming agent and the cleaning agent, a number of the investigations are being performed.

6.2.2. Refrigerant for Heat Pump or Refrigerator

As the alternative freons, HCFC-123, HCFC-141b and HFC-134a are being recommended because of their vapor pressure. The reason why is that their vapor pressures are almost the same as those of CFC-12 and CFC-11, as seen in Fig. 6.2.3 [6.2.3].

Employing these CFCs, a heat pump has been operated and they have been evaluated, as the working fluid for the heat pump. In a performance test wherein the HCFC-123 is employed as the working fluid, a turbo-compression heat pump having a capacity of 180 USRT has been used [6.2.4]. As far as the turbo-compressor is concerned, an external shape of a runner and a width of an impeller are commonly changed according to sizes of a dimensionless head $H_p \cdot g/a^2$ corresponding to an operating temperature condition and dimensionless volume flow $Q/(A \cdot a)$ corresponding to a volume, and it is designed so that an actuating point of the compressor becomes almost the maximum efficiency point. In other words, once the specifications of the runner to be employed are setted, the optimum $H_p \cdot g/a^2$ and $Q/(A \cdot a)$ are accordingly decided (wherein $H_p$ represents a polytropic head, a indicates a suction gas sound speed, Q shows a gas volume flow and A is a cross-sectional area of the sucking portion).

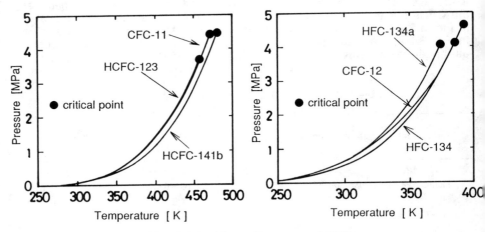

Fig. 6.2.3  Vapor Pressure of CFC

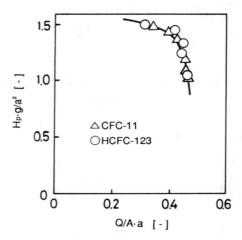

Fig. 6.2.4  Performance of Turbo-Compressor

Consequently, if the $H_p$, a and Q, etc., can be computed according to a physical properties estimating formula, by applying the rule of proportion of the compressor so as to satisfy the similarity conditions specified in JIS-B8354, the performance curves of HCFC-123 and CFC-11 can be plotted on an identical line, thus the performances can be investigated.  As mentioned above, in order to compare the data obtained by operating an identical compressor using the working fluids HCFC-123 and CFC-11, Fig. 6.2.4 has been obtained.  As it is obvious from Fig. 6.2.4, nothing difference in the relationship with $H_p \cdot g/a^2$ and $Q(A \cdot a)$ in both working fluid has been observed. From this fact, the compression efficiency in both case of HCFC-123 and CFC-11 can be identified subject to have an adequate design of the runner of the compressor. For HFC-134a, Hishida et al. [6.2.5] studied on operation of a heat pump consisting of a enclosed rotary compressor. They have found nothing special problem in the operation of the compressor, and the performance of the heat pump was found to be calculated using the thermodynamic properties. It is also the same thing concerning CFC-141b [6.2.6].

As mentioned above, it can be considered that the refrigerant such as CFC-123, etc., being currently proposed are sufficiently usable as the alternatives for the CFC-11, etc.. Moreover, concerning also mixtures of these refrigerant, their adaptability has been investigating. Regarding the thermodynamic properties of the CFC alternatives, Watanabe reviewed in detail [6.2.2].

In operation of compressor, it shall be required to know the influence to machine oil and other organic materials. It is said that the HCFC-123 is soluble in refrigerator oil, and influences the organic materials. The HCFC-123 hardly influences polyamide and melamine resins, however, the high polymer materials such as polycarbonate, polystyrene resins, ABS resin and acrylic resin are dissolved or broken down with the HCFC-123, thus they cannot practically be employed, and it is presumed that also the other materials are greatly influenced with the HCFC-123. When hermetic motor, etc., are employed for the heat pump, the affinity of the refrigerant with the insulating materials shall necessarily be investigated in advance.

In a circumstances of the HFC-134a, it dose not dissolve the refrigerator oil and double-layering phenomenon has been observed in the compressor. Hereafter, further efforts for investigating the most desirable CFC alternatives and the organic materials, shall be expected.

### 6.2.3. Foaming Agent

As the foaming agent for hard and soft foamed products, the CFC-11, CFC-12, CFC-114, CFC113, etc., have been employing. Here, concerning the alternative agents, it would be presented exemplifying a hard polyurethane and polystyrene as the hard foamed products and a soft urethane as the soft foamed product.

Hard polyurethane : In a hard foamed product, the respective foams in the foamed product exist independently, the agent employed as the foaming agent is being included in the independent foams. When the foamed product is employed as a thermal insulator, the thermal conductivity of the agent is desired to be low. It is the matter of course but importance that the agent must be chemically inert to the material subjected to the foaming. As a foaming agent for the hard urethane, the CFC-11 has been used, however in the current situation, the HCFC-123 can be exemplified as an alternative for the CFC-11. The vapor pressure of the HCFC-123 is almost the same as that of the CFC-11 as seen in Fig. 6.2.3, so the foaming mechanisms resemble. However, some problem still exist. For example, the reactivity of

HCFC-123 with the urethane resin is observed. The thermal conductivity of the vapor phase of the HCFC-123 is almost 20% greater than that of the CFC-11. In view of the vapor pressure, the HCFC-141b may be candidate, however, the problem such as reactivity, thermal conductivity, etc., which have seen in the HCFC-123, and in addition to those a new problem of flammability are pointed out. Hereafter, aiming to improve the thermal insulating characteristics by fining the foam and to enhance the strength of foamed product, the detail investigations of reaction conditions including the addition of additives, etc., are desired.

Polystyrene : The CFC-12 has been employed for the polystyrene as a major foaming agent. Historically turning round, at an early stage the foaming has been carried out with LPG, etc., as the foaming agent. From such a historical background, it is said that a part of the industry has already come back to the LPG again, in view of the flammability, the attentions to the manufacturing facilities shall be required in that circumstances. The HCFC-134a and HCFC-22 have been appointed as the alternative agents from their physical characteristics, it is, therefore, said that the replacement in this field is easier than that in the hard urethane foamed product.

Soft Urethane : In the soft urethane foamed product, a majority of foams are produced with the $CO_2$ gas generated by a reaction of the water and isocyanate which are raw materials to produce soft urethane foam, the contribution of the CFC-11 to the reaction is comparatively little accordingly. In another example, a methylene chloride is employed in place of the CFC-11.

### 6.2.4. Cleaning Agent

As the cleaning agent, the CFC-113 has been employing in a broad industrial fields. The reason why is that the CFC-113 has extremely favorable characteristics to apply on the practical industrial processes, that is (1) since the CFC-113 has a small tension which comes from its molecular structure, it soaks into fine clearances of a materials subjected to the cleaning, (2) the solvency of CFC-113 is not so strong as to damage the base materials but sufficiently strong to remove contamination on the base materials, (3) easy handling without toxicity and flammability, and others. The search for the alternative of CFC-113 has been carried out from the same point of view, at present the agent mentioned hereinafter are being proposed.

HCFC-225 : The HCFC-225, the fundamental characteristics of which resemble those of the CFC-113, has been proposed [6.2.7]. The fundamental characteristics of this agent are cited in Table 6.2.2 comparing with those of the CFC-113. It should be said that both of HCFC-225ca and HCFC-225cb are extremely promising because of their boiling point, evaporation rate, KB value and surface tension.

KCD-9434 : KCD-9434 is a nonflammable mixture of HCFC-141b, HCFC-123 and methanol. It has been informed that this agent has low viscosity and a small surface tension [6.2.8].

Pentafluoropropylalchol ("5FP" for short) : It is informed that 5FP has excellent performance in the cleaning of the semiconductor devices using as mixture with a water or a halogenated hydrocarbon according to the purpose of use [6.2.9]. The characteristics of 5FP are also cited in Table 6.2.2.

Table 6.2.2  Physical Properties of HCFC-225ca, HCFC-225cb and 5FP

| | HCFC-225ca | HCFC-225cb | CFC-113 | 5FP |
|---|---|---|---|---|
| Molecular Formula | $CF_2CFC_2HCl_2$ | $CClF_2CF_2CHClF$ | $CCl_2FCClF_2$ | $C_2F_5CH_2OH$ |
| Molecular Weight | 202.94 | 202.94 | 187.38 | 150.05 |
| Boiling Point [°C] | 51.1 | 56.1 | 47.6 | 80.6 |
| Density [g/cm$^3$]@25 °C | 1.55 | 1.56 | 1.56 | 1.50* |
| Viscosity [cps]@25 °C | 0.59 | 0.61 | 0.68 | 2.45* |
| Surface Tension [dyne/cm] | 16.3 | 17.7 | 17.3 | 17.8* |
| KB Value | 34 | 30 | 31 | 36 |
| Flammability | none | none | none | none |

*30 °C

Bioact EC-7 : This is a agent which is novel water soluble solvent [6.2.10] currently being paid attention. The agent consists of a terpene group hydrocarbon contained in a vegetable oil as the major component. Since the terpene is insoluble in the water, it is converted to water soluble by adding a surfactant. In the cleaning of printed circuit board, it is said that its cleaning effect should not be inferior to that of CFC-113.

Another processes such as a water cleaning, etc., have been proposed, however, their details are unknown. Anyway, the research and development of the alternative agent having excellent characteristics of the CFC-113 should be necessary as well as the most suitable cleaning method should be required.

6.2.5. Conclusion

Concerning the alternative refrigerant candidates, HCFC-123, HFC-134a, HCFC-141b and others, their thermodynamic properties have been measured and discussed. The samples of these refrigerant have been provided and the adaptability to the refrigerator or heat pump and the performance as the foaming agent or cleaning agent have been investigating. However, the safety and toxicity have not clarified yet. Presently, programs to confirm the safety are being advanced under the cooperation of the world wide refrigerant manufactures. Now, we are looking forward to hear the conclusion of the programs. Hereafter, the pending problems such as the adaptabilities to organic materials and refrigerator oil shall be investigated in full detail.

Although it has not been mentioned herein, the ammonia is promising candidate as a working fluid for the refrigerator (heat pump). Hereafter, the research and development of technique to safety handle the ammonia overcoming its flammability and toxicity should be expected.

The CFC alternatives mentioned here are generally called by "the second generation freons", and as seen in Table 6.2.1 the ODP or GWP are smaller than those of CFC-11 or CFC-12, but it is not zero. In future, aiming to obtain the ODP, GWP=0, fluorocompound containing oxygen or nitrogen, the cyclic compound containing fluorine and much carbon and branched compound, etc., namely the third generation novel freons shall be developed.

6.2.6. Literature Cited

[6.2.1] Mclinden, M. O. and D. A. Didion: Quest for Alternatives, ASHRAE J. Dec., (1987).
[6.2.2] Watanabe, K.: Current Status of Thermophysical Properties Research on CFC Alternatives, "Heat Pump" ed. by T. Saito and K. Igarashi, Pergamon Press(1990).
[6.2.3] Maezawa, Y. et al.: Measurement of Thermodynamic Properties of HCFC-134 and HCFC-141b, Proceedings of 1989 JAR Annual Conference(1989). (in Japanese)
[6.2.4] Akiya, T. and M. Nakaiwa: Development and Application of CFC-11 Alternative, Chemical Economy (Kagaku Keizai), No.3(1989). (in Japanese)
[6.2.5] Hishida, A. et al.: Results of Operation of Small-Sized Refrigerator Employing HCFC-134a, Proceedings of JAR Annual Conference (1989).
[6.2.6] Blaise, J. C. and T. Dutto: R142b, a Refrigerant for Heat Pumps without Any Damage on Ozone Layer, "Heat Pump" ed. by T. Saito and K. Igarashi, Pergamon Press(1990)
[6.2.7] Yamabe, M.: Establishing Sample Supplying System, Chemical Economy (Kagaku Keizai), No.11(1989). (in Japanese)
[6.2.8] Mitsui-Du pont Fluorochemical Co., Ltd.: Report (1989).
[6.2.9] Daikin Kogyo Co., Ltd.: Report (1989).
[6.2.10] US Patent 4,640,719(1987).

# 6.3 DESTRUCTION TECHNOLOGY OF CFCs

**Tsutomu Sugeta**
National Chemical Laboratory for Industry
Chemical Systems Division, Physico-chemical Properties Section
Agency of Industrial Science and Technology
Ministry of International Trade and Industry
Tsukuba Science City, Ibaraki 305, Japan

## 6.3.1 INTRODUCTION

The usage and the production of CFCs are controlled internationally in accordance with "Montreal Protocol on Substances that Deplete the Ozone Layer" and, furthermore, the Parties adopted further restriction of controlled substance in June of 1990 in order to confirm the protection of the ozone layer. To deal with such situation, developments of alternatives to CFCs are promoted. However, recovery and reuse of CFCs and destruction of waste CFCs are also important in order to control the emission of CFCs until alternatives would be established. In view of conservation of fluorine resources, the recovery/reuse is preferred, but almost all CFCs should be destroyed ultimately for the protection of the ozone layer.

The Protocol defines "production" as "the produced amount of controlled substance minus the amount destroyed by the technologies to be approved by the Parties". In this framework, "the Law concerning the Protection of the Ozone Layer through the Control of Specified Substances and other Measures" stipulates that the competent ministry shall issue standards for the destruction of specified CFCs and halons. Thus the developments of the destruction technologies of the specified substances are urgently necessary. In this section, the present states of the researches and the developements in destruction technologies of CFCs in Japan are mainly summarized.

## 6.3.2 General comments for CFCs destruction

While the usage of specified CFCs in 1986 in Japan is shown in Table 6.3.1, banked CFCs for destruction is not estimated. In U.S., however, the estimated bank is shown in Table 6.3.2[6.3.5]. From these tables, the major candidates for destruction would be CFC-11 and CFC-12, which were used as refrigerants and trapped in rigid foams and CFC-113 used as solvents. It should be noted that they would be wasted as gas or liquid, CFC alone or mixture, etc..

In destruction method under the Protocol, the following matters should be considered [6.3.1].

(1) Reliability
The Protocol allows the Parties to offset CFCs production by their destruction, and reliable destruction methods are required. Therefore, appropriate destruction methods to the souces should be employed.

(2) No environmental impacts
Since highly toxic and hazardous by-product except hydrogen halide might be produced through the decomposition of CFCs, the destruction technologies should not produce them at all or should remove them completely. In particular, in case of the destruction of waste CFCs containing other organic compounds, sufficient care is required.

(3) Low cost
Though an economical aspect should not be prior, the lower the cost of the destruction, the wider the such destruction technology can be adopted.

TABLE 6.3.1 Usage of specified CFCs in Japan in 1986 (Unit:ton).

|         | Refrigerants | Aerosols | Foams  | Solvents | Others | Total  |
|---------|--------------|----------|--------|----------|--------|--------|
| CFC-11  | 2,573        | 4,439    | 21,211 | 305      | 873    | 29,401 |
| CFC-12  | 21,439       | 7,157    | 9,292  |          | 315    | 38,203 |
| CFC-113 | 114          | 159      | 176    | 62,182   | 917    | 63,578 |
| CFC-114 | 134          | 150      | 1,318  |          | 11     | 1,613  |
| CFC-115 | 119          |          |        |          | 11     | 130    |
| Total   | 24,406       | 11,905   | 31,997 | 62,487   | 2,127  | 32,925 |

TABLE 6.3.2 Estimated U.S. bank of CFCs in 1988 (Unit:ton) [6.3.5].

| End Use                     | CFC-11  | CFC-12  | CFC-113 | CFC-114 | CFC-115 |
|-----------------------------|---------|---------|---------|---------|---------|
| Mobile Air Conditioners     |         | 110,000 |         |         |         |
| Other Refrigeration(a)      | 187,200 | 194,130 |         | 1,140   | 16,740  |
| Rigid Polyurethane Foams(b) | 467,000 | 71,600  |         |         |         |
| Other Foams                 |         | 30      |         |         |         |
| Sterilization               |         | 11,500  |         |         |         |
| Solvents                    |         |         | 50,200  |         |         |
| Total                       | 654,200 | 387,260 | 50,200  | 1,140   | 16,740  |

NOTES:
a) Includes: building chillers, home refrigerators & freezers, other refrigeration appliances, refrigerated transport, cold storage warehouse, retail food storage, process refrigeration and liquid food freezing.
b) Excludes flexible polyurethane foam (no banked CFCs).

In the Protocol, the destruction process is defined as "permanent transformation" or "decomposition", and it includes conversion of CFCs into other materials such as HCFCs, HFCs or raw materials of fluorinated plastics. However the "conversion" is not mentioned here.

### 6.3.3 Stability of CFCs and principles of its destruction

Specified CFCs are the compounds substituted hydrogen atoms of methane or ethane by highly electronegative fluorine and chlorine. The various properties characterizing CFCs are owing to the unique properties of the carbon-fluorine bond. As the fluorine has the strongest electronegativity among all elements, it has very strong oxidizing ability. That is, the carbon atom in CFCs is oxidized to its greatest possible extent. For this reason, CFCs destruction by oxidation, which is a conventional treatment of organic waste, is difficult. Comparison of the change of enthalpy among typical reactions for CFCs destruction are shown as follows.

OXIDATION:
$CCl_2F_2 + O_2 = CO_2 + Cl_2 + F_2$; $\Delta H= +83.3 kJ/mol$ (1)
HYDROLYSIS:
$CCl_2F_2 + 2H_2O = CO_2 + 2HCl + 2HF$; $\Delta H=-160.2 kJ/mol$ (2)
INCINERATION:
$CCl_2F_2 + CH_4 + 2O_2 = 2CO_2 + 2HCl + 2HF$; $\Delta H=-959.4 kJ/mol$ (3)

It is found that oxidation of CFCs is difficult, on the other hand, hydrolysis or incineration with fuel can be effective.

### 6.3.4 Destruction technologies of CFCs

Since it has been believed that CFCs are very stable and harmless until a recent date, there has been no demands for CFCs destruction and few researches of destruction/decomposition of CFCs. Recently some technologies for the destruction of CFCs have been proposed also in Japan [6.3.1,2].

(1) Incineration and thermal decomposition
CFCs can be destroyed by incineration with fuel as shown in Eq.(3). It is thought that this reaction is a kind of hydrolysis caused by water produced through combustion of hydrocarbon.

Tokuhashi et al. [6.3.3] report that CFC-12 is destroyed efficiently by pre-mixed combustion of CFC-12/methane/air using an experimental apparatus shown in Fig.6.3.1. In this apparatus, diameter of the burner nozzle is 9.6 mm, and CFC-12, methane and air, whose flow rates are controlled by thermal mass flowmeters, are mixed perfectly and burned. The CFC remained in combustion gas are analyzed by gas chromatograph. Figure 6.3.2 shows the change of decomposition yield with addition of CFC-12 to stoichiometric mixture of methane/air. When the ratio of CFC-12 to methane is lower than 0.2, the concentration of CFC-12 in combustion gas are below the detection limit of gas chromatograph and almost all CFC-12 are decomposed. The decomposition yield reduces with increase in the ratio of CFC-12 to methane and is about 0.92 at the ratio of 0.53.

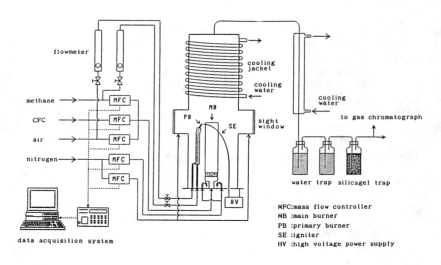

FIGURE 6.3.1 Experimental apparatus for CFC destruction through incineration.

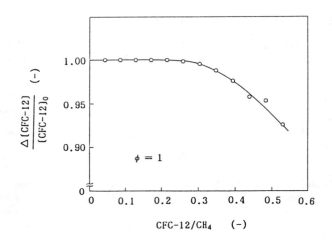

FIGURE 6.3.2 Relation between decomposition yield of CFC-12 and ratio CFC/CH$_4$.

Graham et al. [6.3.4] report thermal degradation of the hazardous waste mixture consisted of, by weight, 2.5% chlorobenzene, 2.5% tetrachlorocarbon, 2.5% trichloroethylene, 2.5% CFC-113 and 90% toluene by using the thermal decomposition unit-gas chromatographic system. The reaction conditions are in excess, deficient and stoichiometric oxygen with respect to complete combustion. Under the stoichiometric combustion condition, the temperature required for 99% destruction of CFC-113 at a gas-phase residence time of 2.0 sec is 1053 K. Many kinds of thermal reaction products are observed for every case of the three reaction conditions.

The incineration is considered to be employed for destruction of CFCs at commercial scale most immediately. A trial burn in U.S. shows a destruction efficiency above 99.9998 % for co-fired CFC-11 and CFC-12 [6.3.5], but the details are not reported. However, the formation of toxic and hazardous by-products would be concerned, especially, in case of co-incineration with other hazardous waste or some kinds of supplemental fuels. Therefore, an optimum incineration condition and process design must be examined thoroughly as well as the treatment of combustion flue gas.

(2) Decomposition with supercritical water
The properties of supercritical water above its critical temperature (647.3K) and pressure (22.12 MPa) lie between those of liquid water and steam, and they are dependent strongly on the density. Supercritical water near the critical temperature has liquid-like characteristics. For example, the ion product is nearly equal to or greater than that of liquid water at room temperature, and the dielectric constant is comparable to that of a polar organic liquid. Therefore hydrolysis is preferable to pyrolysis in supercritical water. Supercritical water is a dense gas in which organic compounds such as CFCs are completely miscible. Therefore supercritical water would decompose CFCs easily.

Modell [6.3.6] describes an efficient processing method for the oxidation of organic materials containing chlorinated hydrocarbons in supercritical water. However the applicability to CFCs destruction is not reported.

The authors [6.3.7] carried out an introductory investigation of a hydrolysis of CFCs in supercritical water. The CFC and water are loaded into microreactor and its temperature is held at 773 K for 45 minutes. Then hydrogen chloride and hydrogen fluoride produced are analyzed. Fig.6.3.3 shows the effect of water density on the yield of hydrogen halides in hydrolysis of CFC-11 and CFC-113. While dechlorination is almost completed above the density of 0.2 g/cm$^3$ in both CFCs, defluorination is rather difficult particularly in case of CFC-113 which is decomposed at the density of 0.5 g/cm$^3$.

Since a great amount of water is used as the reaction medium in this method, toxic by-products except hydrogen halides would be hardly produced. The hydrogen halides are easily trapped in aqueous phase and become harmless by neutralization. Therefore, a closed process can be realized. On the other hand, as the plant is exposed to corrosive condition, the selection of plant materials is important.

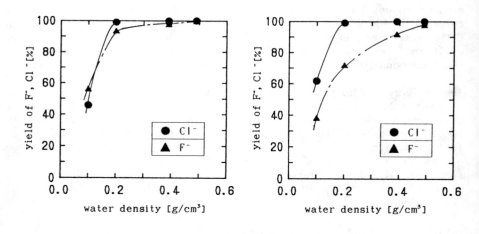

(a) CFC-11   (b) CFC-113
FIGURE 6.3.3 Relation between yield of hologen ion and water density through supercritical water hydrolysis of CFC.

(3) Catalytic decomposition
In this method, CFC is decomposed catalytically by hydrolysis rather than hydrogenation and disproportionation. CFC and water, accompanied by carrier gas, is fed on solid catalyst and hydrolyzed.

Tajima et al. [6.3.8] report that Y-type, mordenite-type and ZSM-5-type zeolites show effective activity for decomposition of CFC-113 as shown in Table 6.3.3.

TABLE 6.3.3 Catalyst activity for decomposition of CFC-113.
(at 773 K)

| Catalyst | decomposition yield(%) |
| --- | --- |
| HY-type zeolite | 85 |
| NaY-type zeolite | 15 |
| mordenite-type zeolite | 98 |
| ZSM-5-type zeolite | 88 |
| $\gamma$-$Al_2O_3$ | 85 |
| $SiO_2TiO_2$ | 40 |

Okazaki et al. [6.3.9] report decomposition of CFCs by ferric oxide catalyst supported on activated charcoal. The order of reactivity of CFCs is as follows:

$$CCl_4 > CCl_3F > CCl_2F_2 > CClF_3 > CF_4$$

All materials except tetrafluoromethane are decomposed almost completely from 673 K to 863 K, but tetrafluoromethane is scarcely decomposed even at 863 K. This order is consistent with that of strength of C-Cl and C-F bonds. The reactivity of CFC-113 is between CFC-11 and CFC-12.

This method might be appropriate to the industrial waste gas containing dilute CFC, because air can be used as carrier gas [6.3.8]. The problems of this method are developments of long-life and highly active catalyst.

(4) Plasma decomposition

Wakabayashi et al. [6.3.10] report CFC decomposition using a high frequency-induced plasma reactor. Thermal decomposition or/and hydrolysis occur in high temperature plasma reactor (about 10000 K). When CFC is reacted alone, soot and various products by disproportionation and dimerization are formed. But adding water prevents their formation and CFC is hydrolyzed. Though this method would treat CFCs with high speed, it has the problems for the treatment of a large amount of waste gas and a high cost of electricity.

(5) Chemical reductive method

As a destruction method of CFCs by dehalogenation via reducing chemicals, Oku et al. [6.3.11] report a method using sodium naphthalenide dissolved in organic solvent. When CFC-113 is reacted with 1.5 equivalents of naphthalenide and an additive in THF at 423 K for 50 min, almost complete dehalogenation is achieved. This method is carried out at lower temperature than that of the other methods, but high reagent costs are required.

(6) Others

In addition of the methods mentioned above, the following methods are listed in reference[6.3.5].
  microbial attack
  active metals scrubbing
  highly alkaline scrubbing
  corona discharge

## 6.3.5 Conclusion

Though the destruction technologies of CFCs must be urgently established, most researches have been started recently and there are no technologies to put into practical use immediately. Incineration would be established on commercial scale in a near future, but various technologies should be also prepared to apply to every kind of waste CFCs. There is concern for formation of toxic by-products through the decomposition of CFCs and, therefore, the destruction conditions and the standards of assessment of the destruction results should be defined in consideration of environmental impacts.

Most destruction technologies of CFCs would be applicable to destruction of CFC alternatives (HCFCs) which may be controlled in the next century.

**REFERENCES**

[6.3.1] Ministry of International Trade and Industry, Destruction Technologies of CFCs (Interim Report), April 1989.

[6.3.2] Committee for Protection of the Stratospheric Ozone Layer, Destruction Technology of CFC's - Second Interim Report of the Working Group for Emissions control and Destruction Technology, January 1990.

[6.3.3] Tokuhashi, K., et al., 22th Symp. Safety Eng., Japan, p.30, 1989.

[6.3.4] Graham, J. L., Hall, D. L. and Dellinger, B., Environ. Sci. Technol., vol.20, p.703, 1986.

[6.3.5] Dickerman, J. C., Emmel, T. E., Harris, G. E. and Hummel, K. E., Final Report Technologies of CFC/Halon Destruction, EPA 68-02-286, April 1989.

[6.3.6] Modell, M., U.S. Patent 4338199(1982) and 4543190(1985).

[6.3.7] Sugeta, T., et al., 22th Autumn Meeting of Soc. Chem. Engrs., Japan, SL104, 1989.

[6.3.8] Tajima, M., et al., 30th Annual Meeting of Japanese Soc. of Air Pollution, 7321, 1989.

[6.3.9] Okazaki, S., et al., 64th Symp. Catalyst, Japan, 3C30, 1989.

[6.3.10] Wakabayashi, T., et al., 9th Int. Symp. Plasma Chem., Sept. 4-8, 1989.

[6.3.11] Oku, A., Kimura, K. and Sato. M., Chem. Lett., 1789, 1988.

# 7. NOx, SOx AND ACID RAIN PROBLEMS

**Taeko Sano**
Tokai University
Faculty of Engineering
1117 Kitakaname, Hiratsuka, Kanagawa 259-12, Japan

7.1 Present Status of Air Pollution in Japan

In Japan, environmental quality standards [7.1] for $NO_2$ (nitrogen dioxide) and $SO_2$ (sulfur dioxide) are established. It requests for $NO_2$ that "daily average of hourly values shall be no more than 0.04 ppm" and that "in an area where the daily average is between 0.04 ppm and 0.06 ppm, efforts should be made so that the ambient concentration is maintained around the present level or does not significantly exceed the present level". It also requests for $SO_2$ that "the daily average of hourly values must be less than 0.04 ppm and the hourly values less than 0.1 ppm". In order to monitor the state of the compliance with environmental quality standards, monitoring stations are located and the concentrations of $NO_2$ and $SO_2$ are recorded. In 1987, the data of $NO_2$ concentration in ambient air were collected at 1321 valid monitoring stations, where monitoring was conducted for more than 6000 hours a year and for automobile exhaust at 282 valid automobile exhaust monitoring stations, where the sampling inlets were placed at the outside of vehicle lanes, and then the data of $SO_2$ concentration in ambient air at 1608 valid monitoring stations.
Figure 7.1 shows changes in the annual average concentrations of $NO_2$ which have been measured respectively at 15 ambient air pollution and 26

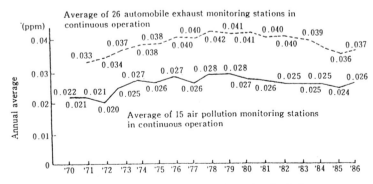

FIGURE 7.1 Annual Average Concentration of $NO_2$ [7.1].

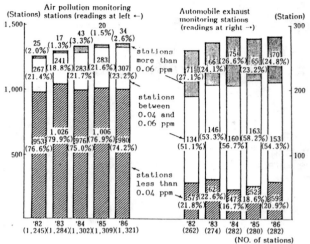

Notes: 1. The automobil exhaust monitoring stations do not include those the sampling inlets are located in the driveway.
2. The figures in parentheses indicate the ratio of number of the given monitoring stations to number of all pollution monitoring stations.
3. The concentration level is annual 98% value of daily average.

FIGURE 7.2 Compliance Status with Environmental Quality Standards [7.1].

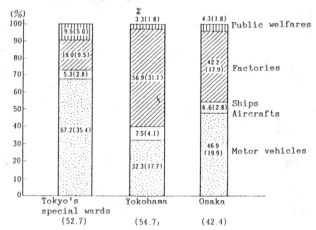

Note  1. Estimated by Environment Agency.
2. ( ):the amount of NOx exhausted (thousand t/year).

FIGURE 7.3 NOx Emissions in the Area for Areawide Total Pollutant Load Control [7.1].

automobile exhaust monitoring stations in continuous operation. $NO_2$ concentrations recorded take a peak value of 0.028 ppm at air pollution monitoring station and of 0.042 ppm at automobile exhaust monitoring station in 1978, and then show a decreasing tendency. However, they

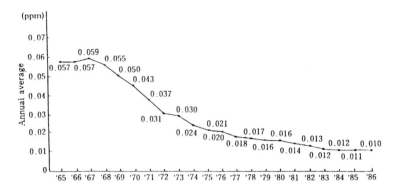

Note: Annual average in the quotient of the sum of all one-hour values per year divided by the total measured hours.

FIGURE 7.4 Annual Average Concentration of $SO_2$ [7.1].
(Average of 15 air pollution monitoring stations)

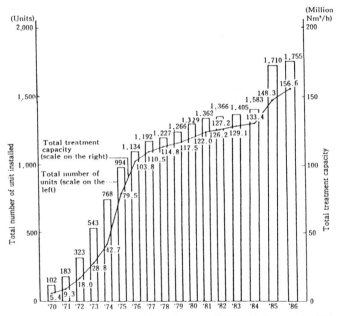

Note: Figures are as of January 1 for each year until 1983, and the ones after 1983 are as of March 31 of each year.

FIGURE 7.5 Number and Total Capacities of Flue Gas Desulfurization Units [7.1].

increase again from 1985 to 1986. Figure 7.2 shows the compliance status with environmental quality standards for $NO_2$. In 1986, monitoring stations where $NO_2$ concentration exceeds the upper limit of the environmental quality standards, 0.06 ppm, are 24.8 % for automobile exhaust monitoring and 2.6 % for air pollution monitoring. Figure 7.3 shows the sources of nitrogen oxides (NOx=NO+NO$_2$) at Tokyo, Yokohama and Osaka in 1985. A large portion of nitrogen oxides is emitted from combustors of industries (stationary sources) and automobiles.

Figure 7.4 shows changes in the annual average concentration of SO$_2$ which has been recorded at 15 air pollution monitoring stations since 1965. The level of SO$_2$ has decreased from the peak value of 0.059 ppm in 1967 to 0.01 ppm in 1986. In 1986, at 98.9 % of air pollution monitoring stations, SO$_2$ concentration did not exceed the environmental quality standard. Although sulfur oxides (SOx=SO$_2$+SO$_3$) in ambient air are also generated by combustion of hydrocarbon fuels such as coal and petroleum with sulfur, good compliance with the environmental quality standard for SO$_2$ is attained by installing the fuel and flue gas desulfurization units. Figure 7.5 shows changes in the number and total capacities of flue gas desulfurization units.

Acid rain is considered to be formed through complex photochemical reactions of NOx, SOx and hydrocarbon, and absorption by particulates of rain, cloud and fog. Acid rain tends to descend in area several kilometers from their origin. Japan recently started out to investigate the influence of acid rain on the forest and woods.

## 7.2 Formation and Control of NOx

In the combustion processes of hydrocarbon fuels, nitric oxide (NO) is produced predominantly in practical combustors and exhausted into the atmosphere, where NO is oxidized to NO$_2$ by photochemical reactions. It is also reported, however, that a larger portion of NOx consists of NO$_2$ in the gas turbine combustor [7.2] and further, that the unvented gas-fired heaters tend to emit NO$_2$ over the environmental quality standard [7.3]. However, since NO$_2$ is generated by oxidation of NO in the combustors [7.4,7.5], the formation and control of NOx in combustors should be reduced to those of NO in combustors.

### 7.2.1 Formation Mechanism of NO

In the combustion processes of hydrocarbon fuels with nitrogen compounds (fuel N), the nitrogen compounds in fuels are oxidized to nitric oxide, so called fuel NO, and its formation mechanism is considered to take the process through NH$_i$ (i=0,1,2) as follows.

$$\text{fuel N} \underset{\text{HCN}}{\rightleftarrows} \text{NH}_i(i=0,1,2) \begin{array}{c} +\text{OH} \rightarrow \text{NO} \\ +\text{NO} \rightarrow \text{N}_2 \end{array} \quad (7.1)$$

In the combustion of hydrocarbon fuels without fuel N, nitrogen (N$_2$) in air is also oxidized to nitric oxide at high combustion temperatures. It is called as thermal NO. The formation rate and mechanism of thermal NO generated in the flame is different from those generated in the post flame zone. The former is called as prompt NO and its formation rate is much more rapid than that of the latter. The formation mechanism of prompt NO is proposed as follows.

$$CH_2 + N_2 = HCN + NH$$

$$CH + N_2 = HCN + N$$

$$HCN + OH = CN + H_2O$$

$$CN + O_2 = CO + NO \qquad (7.2)$$

$$CN + O = CO + N$$

$$NH + OH = N + H_2O$$

The formation rate of thermal NO generated in the post flame zone is fairly slow and can be predicted by the reactions (Eq.(7.3)) proposed by Zeldovich with the reaction of Eq.(7.4), and it is usually called as Zeldovich NO.

$$\left. \begin{array}{l} N + O_2 = NO + O \\ O + N_2 = NO + N \end{array} \right\} \qquad (7.3)$$

$$N + OH = NO + H \qquad (7.4)$$

The characteristic properties of NO formation in the combustion processes can be summarized as follows.

(1) Fuel NO       : Formation by oxidation of nitrogen compounds (fuel N) in fuels.

(2) Thermal NO    : Formation by oxidation of nitrogen ($N_2$) in air at high combustion temperatures.

  (i) Prompt NO    : Formation in the flame - rapid rate and small amount of formation.

  (ii) Zeldovich NO : Formation in the post flame zone - slow rate and large amount of formation, which depends on
(a) combustion (flame) temperature,
(b) residence time, and
(c) oxygen concentration.

## 7.2.2 Control of NO

In order to reduce the formation of fuel NO, first of all, the use of good quality fuels containing no fuel N is most desirable. However, low quality oil and coal which contain fuel N are practically used for conventional combustors for boilers of power station. Table 7.1 shows the content of fuel N in fuels, and Table 7.2 shows the contribution of thermal NOx and fuel NOx to NOx emissions of gas, oil and coal fuels [7.6]. Figure 7.6 shows the features of prompt NO formation in premixed propane-air-$O_2$ flat flames at atmospheric pressure [7.7]. The maximum flame temperature is controlled to be constant for different equivalence ratios. For the equivalence ratio $\phi \leq 1.0$, the formation of prompt NO is small and Zeldovich NO is the main source of NO. Although prompt NO

TABLE 7.1 Contents of Fuel N [7.6].

| Fuel | Contents % |
|---|---|
| Crude Oil(The Middle East) | 0.09 ~ 0.22 |
| Heavy Oil C | 0.1 ~ 0.4 |
| Heavy Oil A | 0.05 ~ 0.1 |
| Light Oil | 0.002 ~ 0.03 |
| Kerosene | 0.0001 ~ 0.0005 |
| Coal | 0.2 ~ 3.4 |
| LPG, City Gas | 0 |

TABLE 7.2 Thermal NOx and Fuel NOx [7.6].

| Fuel | Gas | Oil | Coal |
|---|---|---|---|
| Thermal NOx | 100 % | 30   40 % | 10   20 % |
| Fuel NOx | 0 % | 60   70 % | 80   90 % |

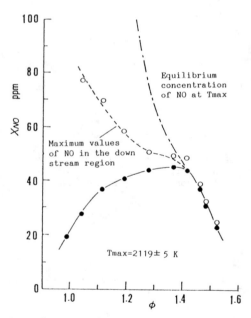

FIGURE 7.6 Formation Features of Prompt NO [7.7].
($C_3H_8$/air+$O_2$, atmospheric flat flame)

is considerably formed for $1.0 < \phi \leq 1.4$, the concentration is lower than the equilibrium concentration at the maximum flame temperature, and Zeldovich NO is still formed in the post flame zone. For $\phi > 1.4$ the formation of Zeldovich NO in the post flame zone is not found.

In order to reduce NOx emission in practical combustors, of which the equivalence ratio is usually less than 1.0, it is most important to develop technologies for controlling the formation of Zeldovich NO in the post flame zone. However, it has been recently reported for low NOx combustion developed for pulverized coal that the contribution of prompt NO to NOx emission would not be neglected [7.8].

Based on the above-mentioned fundamental knowledges of formation and reduction of fuel NO and thermal NO in combustion processes, the control methods of NOx emission can be summarized as follows.

(1) Use of good quality fuels,

(2) Combustion control of NO formation by

    (i) Reduction in combustion temperature with
        (a) Exhaust (flue) gas recirculation
        (b) Lean mixture combustion
        (c) Water (steam) injection
        (d) Catalytic combustion
        (e) Use of emulsified fuel
        (f) Low compression-ratio combustion

    (ii) Combustion in short residence time with
        (a) Well mixing
        (b) Optimization of ignition time
        (c) Improvement of position and number of ignition plugs
        (d) Optimization of fuel injection process
        (e) Improvement of combustion chamber shape

    (iii) Combustion in low oxygen circumstance (in reductive circumstance) with
        (a) Two stage combustion
        (b) Fuel injection (in-furnace NOx reduction)

(3) Exhaust (Flue) gas denitrification by

    (i) Catalytic method with
        (a) Non-selective catalyst (three-way catalyst)
        (b) Selective catalyst

    (ii) Non-catalytic method with
        (a) Reduction by ammonia
        (b) Electron irradiation

By applying and combining these methods to reduce NOx emissions from combustion, various types of low NOx burners have been developed, and used successfully in practice. The two stage combustion technology was originally developed to control the formation of Zeldovich NO. In the primary combustion zone, fuels burn in the low oxygen circumstance (in fuel rich flame) to result in low NO formation, and then, by introducing secondary air at low combustion temperature, the unburned hydrocarbons are oxidized to complete combustion process. The two stage combustion technology is effective in reducing fuel NO, because fuel NO is diminished by reduction reaction with the intermediate chemical species such as $NH_i$ (i=0,1,2), CN, et al., which are produced in the fuel rich combustion;

$$NH + NO = N_2 + OH$$
$$CN + NO = N_2 + CO \qquad (7.5)$$
$$NH_2 + NO = N_2 + H_2O$$

Therefore, the two stage combustion technology has been widely applied for reducing fuel NO as well as thermal NO, although the most important point is how to introduce the secondary air in the combustion process. The fuel injection method (in-furnace NOx reduction) has been also developed to reduce effectively NOx emission, in which fuel is further injected in the down stream of the primary combustion zone to promote NOx reduction.

7.2.3 NOx Emission Control from Stationary Sources : Boilers, Furnaces

In Japan, national emission standards [7.1] for NOx from stationary sources were established in August 1973. Since then they have been tightened and the types of facilities to be subject to the regulation have been increased four times. In September 1983, the emission standards for solid fuel combustion boilers were made tighter, and in June 1985, the emission control was stretched to small boilers. The emission standard levels are established concerning the fuels, the type of facilities, the stack gas volume and the date of installation.

In order to reduce NOx emissions from boilers, combustion control technologies applied are two stage combustion, flue gas recirculation and in-furnace NOx reduction (fuel injection), flue gas denitrification, and further their combinations. Figure 7.7 shows the effect of flue gas recirculation on NOx emissions [7.6]. It is seen from the figure that the flue gas recirculation could be very effective for the gas and oil fired boilers, although its effect decreases with the increase in the content of fuel N. For coal fired boilers, the flue gas recirculation is not shown to be effective.

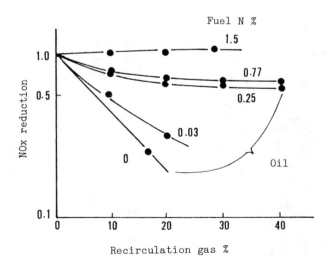

FIGURE 7.7 Effect of Flue Gas Recirculation on NOx Reduction [7.6].

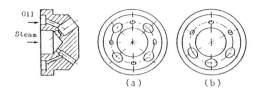

FIGURE 7.8 Arrangement of Fuel Injection Nozzles for Low Nox Burner [7.6].

Recently, various types of low NOx burner have been developed and successfully reduced NOx emissions. The concept of low NOx burners for boilers and furnaces is that NOx emission is reduced by controlling combustion in the vicinity of the burner; that is, (1) by controlling the formation of thermal NO by lowering flame temperature with changes in the size and arrangement of fuel injection nozzles and injection timing, and then (2) by reducing generated NO, which includes fuel NO, by reactions with combustion intermediate species. Figure 7.8 shows the arrangement of burner nozzles developed for an oil fired furnace [7.7]. Figure 7.9 shows the concept of a low NOx burner for pulverized coal combustion developed by Hitachi Co., Ltd [7.9].

In region A, volatile nitrogen compounds generated by thermal decomposition of pulverized coal are oxidized to fuel NO, and in region B they react with combustion intermediate species such as CH and $C_2$ and reducing agents such as $NH_i$ and CN are produced. In region C, thermal NO is generated by combustion of char, and in region D, NO is reduced by reactions with reducing agents. Char C is also a reducing agent for NO. By using the combination technologies of the above mentioned low NOx burner and in-furnace combustion, the NOx emission from the pulverized coal fired boiler has attained to the level less than 150 ppm (6% $O_2$).

Figure 7.10 shows the annual trend and state of the establishment of flue gas denitrification units. The majority of these units makes use of the process of dry selective catalytic reduction, while others use non-catalytic reduction process, wet-type direct absorption and wet-type

Region A:
Volatile N + $O_2 \rightarrow$ NO

Region B:
Fuel + $O_2 \rightarrow$ CH, $C_2$, etc.
CH + NO $\rightarrow NH_i$ + CO
$C_2$ + NO $\rightarrow$ CN + CO
Volatile N $\rightarrow NH_i$, CN

Region C:
Char N + $O_2 \rightarrow$ NO

Region D:
NH + NO $\rightarrow N_2$ + OH
CN + NO $\rightarrow N_2$ + CO
Char C + NO $\rightarrow N_2$ + CO

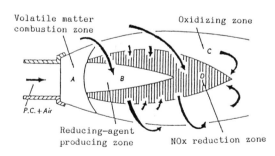

FIGURE 7.9 The Concept of a Low NOx Burner for Pulverized Coal Combustion [7.9].

oxidation absorption [7.1]. By using the unit with 80 % denitrification capacity, NOx emission can be reduced to levels of 10-20 ppm for gas firing, 20-30 ppm for oil firing and about 60 ppm for coal firing [7.6].

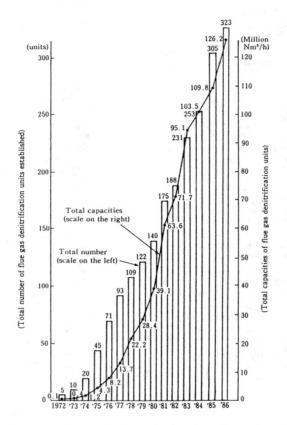

Notes: 1. Figures for 1984 and after are based on the projects performed for preparation of the data reported about the facilities that emit flue gas.
2. Figures are as of January and each year nutil 1982, fsi the ones after 1983 are as of murh 31 of each year.

FIGURE 7.10 Annual Trend and State of Establishment of Flue Gas Denitrification Units [7.1].


7.2.4 Stationary Internal Engine

The emission standards for gas turbine and diesel engine were established in October 1987. and national emission standard levels were legislated in December 1988. Tokyo metropolitan emission standard levels were enacted in January 1989 for gas engines as well as gas turbine and diesel engine, being much more severe than those of the national emission standards.

(1) Gas turbines

To reduce NOx emission from gas turbine combustors, (1) use of good quality fuels, (2) combustion control: premixed combustion(lean mixture combustion), water injection etc., and (3) flue gas denitrification have been applied.

(i) Low NOx combustor

Figure 7.11 shows a low NOx gas turbine combustor developed by Mitsubishi Heavy Industries, Ltd.[7.10], which is applied for the 120 MW gas turbine of a 1090 MW LNG combined cycle plant. The combustor consists of a pilot stage with a conventional diffusion nozzle and a main stage with a premixing nozzle. At low engine power operations and starting with low fuel flow rates, only the pilot stage is fueled and the air-bypass valve is closed. Just before the main stage is fueled at around 38 percent engine power conditions, the bypass valve is fully opened. Then 35 percent of combustion air bypasses the pilot and main stages, and flows directly into the upstream end of the transition piece. Above this point, the main fuel is injected and controlled in the scheduled ratio of the main to pilot fuel flow rate. As the engine power increases, the air-bypass valve is gradually closed to keep the fuel-to-air ratio constant in both pilot and main stages, and is almost closed at the base load. In the main stage, the formation of NO is controlled to be low due to premixed lean mixture combustion. Figure 7.12 shows the NOx emission versus the turbine inlet temperature. The NOx emissions were corrected as to correspond specific humidity of 0.008 kg $H_2O$/kg dry air. The NOx emissions are lower than the target value of 75 ppm, and the CO and unburned hydrocarbon emissions are approximately 34 ppm and 8 ppm at base load conditions, respectively. The best premixed equivalence ratio is at around 0.7 over the whole operating range while a main stage is fueled.

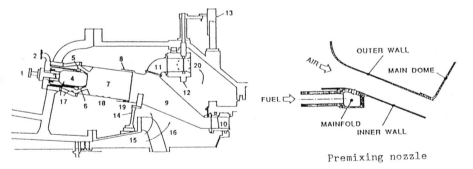

| | | | |
|---|---|---|---|
| 1 Pilot fuel delivery | 6 Premixing nozzle | 11 Bypass elbow | 16 Compressor discharge air |
| 2 Main fuel delivery | 7 Main stage | 12 Butterfly valve | 17 Combustion air |
| 3 Pilot fuel nozzle | 8 Clam shell | 13 External ring | 18 Cooling air |
| 4 Pilot stage | 9 Transition piece | 14 Flexible support | 19 Dilution air |
| 5 Main fuel nozzle | 10 Turbine nozzle | 15 Diffuser | 20 Bypassing air |

FIGURE 7.11 Low NOx Gas Turbine Combustor [7.10].

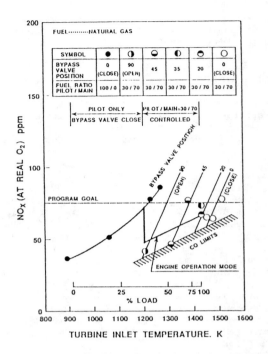

FIGURE 7.12 NOx Concentration Emitted from Low NOx Gas Turbine Combustor [7.10].

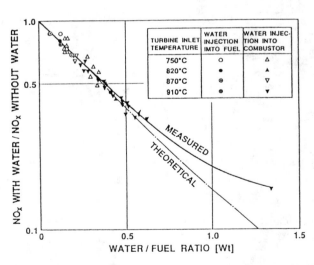

FIGURE 7.13 Effect of Water Injection to NOx Emission [7.11].

Figure 7.13 shows the effect of water injection to NOx emissions for a 60 MW gas turbine. Both water injections into the combustor through atomizing air holes (Fig.7.14-a) and into the fuel passage before the nozzle (Fig.7.14-b) are found to have little difference in NOx emission [7.11].

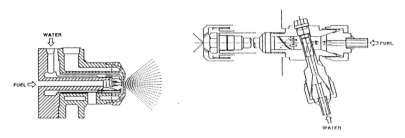

(a) Direct Injection into the Combuster

(b) Mixing with Fuel

FIGURE 7.14 Water Injection Methods.

(2) Diesel engines

To reduce NOx emission from stationary diesel engines, the applicable methods are based on (1) the change in fuel: the use of high cetane fuels, good quality fuels (without fuel N) and emulsified fuels, (2) the combustion control technologies: optimization of the fuel injection time, water injection, exhaust gas recirculation, improvement of the fuel injection system, pilot injection, improvement of the combustion chamber shape, decrease in the compression ratio, use of indirect diesel engines

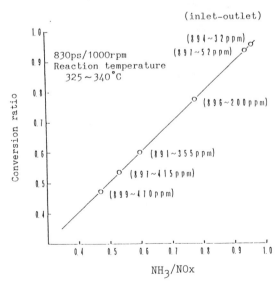

FIGURE 7.15 The Conversion Efficiency of NOx [7.12].

and (3) the exhaust gas denitrification. However, NOx emissions were reduced only from 30 % to 60 % by combining the change in fuels and the combustion control technologies. Therefore, in order to attain the emission standard level, the exhaust gas denitrification technology must be applied. The dry selective catalytic reduction process with ammonia ($NH_3$) is one of the most effective methods, because of involving about 13 % oxygen in exhausted gases.

Figure 7.15 shows the conversion behavior of NOx emission with ammonia for the exhausted gas from direct injection diesel engines by using a denitrification facility developed by Niigata Engineering Co.,Ltd.[7.12].

The dry selective catalytic reduction process with ammonia is not influenced by the oxygen concentration in the exhaust gas, but requests oxygen to activate the reduction point of catalyst. NOx in the exhaust gas is reduced by the following reactions

$$4\ NO + 4\ NH_3 + O_2 \longrightarrow 4\ N_2 + 6\ H_2O$$
$$NO + NO_2 + 2\ NH_3 \longrightarrow 2\ N_2 + 3\ H_2O$$
(7.6)

The most adequate temperatures for the catalytic reduction ranges from 300°C to 400°C. The exhausted gases from diesel engines have the temperatures in this range. In Fig.7.15, the catalyst of $WO_3$-$V_2O_5$-$TiO_2$ has been used. The effect of denitrification reaches a maximum at $NH_3/NOx=1$. The quantity of ammonia to be injected in the denitrification process can be determined by monitoring the NOx concentration in the exhausted gas from diesel engines.

7.2.6 NOx Emission Control from Motor Vehicles

The regulation for NOx emissions from motor vehicles have been enacted since 1973 for gasoline- and LPG-fueled vehicles and since 1974 for diesel-power vehicles. Subsequently, for gasoline- and LPG-fueled passenger cars, regulations based on the initial target standards(average emission of NOx: 0.25g/km) were enforced in 1978 (1978 regulation). Regulations on automobiles other than gasoline-and LPG-fueled passenger cars (trucks, buses and so fourth) were started in 1973-74 and were further strengthened in 1975 and 1977. After that, the first phase regulations on all types of vehicles were carried into effect in 1979 and the second phase regulations had been enforced on all types of vehicles

TABLE 7.3 Targets of NOx Reduction from Diesel-powered Passenger Cars (Average) [7.1].

| Type | 1st phase target | | 2nd phase target | Method of measurement |
|---|---|---|---|---|
| | Target value for regulation (average) | Tine for application of the regulation | | |
| Small-sized vehicle (Equivalent inertia weight less than 1,250 kg) | 0.7 g/km | From October 11, 1986 for vehicles with manual transmission. | 0.5 g/km | 10-mode |
| Medium-sized vehicle (Equivalent inertia weight exceeding 1,250 kg) | 0.9 g/km | From October 11, 1987 for vehicles with automatic transmission | 0.6 g/km | |

since 1983 [7.1]. Diesel-powered passenger cars have been subjected to two-step reduction targets. The first phase regulations which were carried into effect in 1986 and 1987, and the second phase regulations to be made effective in 1990 and 1992 are shown in Table 7.3 [7.1].

(1) Gasoline fueled vehicles
At present, the NOx emission controls from gasoline fueled vehicles are generally carried out by the catalytic denitrification of exhaust

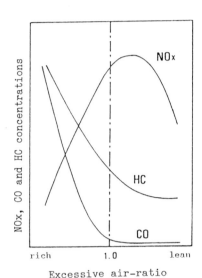

FIGURE 7.16 Concentrations of NOx, CO and HC in Exhaust Gases [7.13].

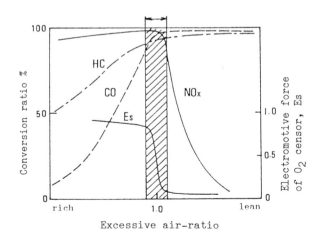

FIGURE 7.17 Conversion Ratio of NOx, CO and HC by a Three-way Catalyst [7.14].

gases, especially, the simultaneous removals of NOx, carbon monoxide (CO) and unburned hydrocarbon (HC) with three-way catalysts. Figure 7.16 shows the concentrations of NOx, CO and HC in exhaust gases versus the excessive air-ratio [7.13]. As a function of the excessive air-ratio, the emission features of NOx are opposite to those of CO and HC. Figure 7.17 shows the conversion efficiency of NOx, CO and HC versus the excessive air-ratio by a three-way catalyst, which is shown to be the most effective for NOx, CO and HC near the stoichiometric air-fuel ratio [7.14]. In order to control suitably the air-fuel ratio of exhaust gases, the oxygen concentration in the exhaust gas should be measured by an $O_2$ censor, of which electromotive force changes stepwisely near the stoichiometric air-fuel ratio. By applying the signal from the $O_2$ censor with measuring the intake air flow rate and others, the opening and shut-off times for the electromagnetic valve of fuel injection are regulated to maintain the stoichiometric air-fuel ratio. Figure 7.18 shows such an emission control system developed by Toyota Motor Co. [7.15].

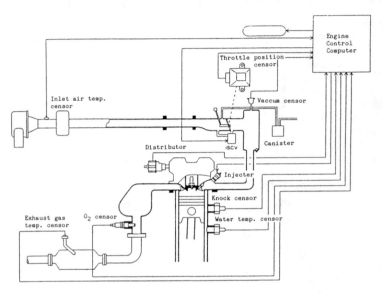

FIGURE 7.18 The Emission Control System [7.15].

There are two types of the catalyst, that is, monolith and pellet. The former is mainly used because of better performance of "warm up" and lower ventilation resistance. The monolith catalyst as shown in Fig.7.19 has a honeycomb structure, which is made of ceramic substrate (cordierite:$2MgO-2Al_2O_3-5SiO_2$) with noble particles such as platinum (Pt) and rhodium (Rh) of sizes less than 10 Å. To stabilize the alumina support, and to improve the thermal durability and catalytic activity, ceria ($CeO_2$) and lanthanum (La) are added [7.16]. Ceria absorbs the variation of oxygen concentrations in the exhaust gases by the following reaction

$$CeO_2 = CeO_{2-m} + 0.5 \, m \, O_2 \qquad (7.7)$$

The reduction efficiency of the Rh catalyst is improved significantly

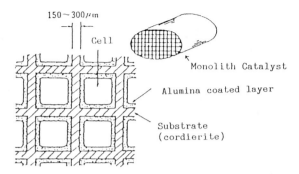

FIGURE 7.19 Monolith Catalyst [7.16].

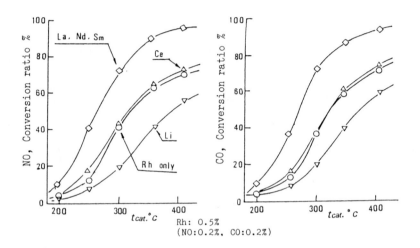

FIGURE 7.20 Effect of Lanthanum (La) to Rhodium (Rh) Catalyst.

by addition of La (Fig.7.20). The catalytic reaction of NO with CO on the Rh catalyst is given by

$$2\ NO + 2\ CO \longrightarrow 2\ CO_2 + N_2 \tag{7.8}$$

At starting of the engine, the temperature of the catalyst is lower than that necessary for the catalytic reaction. Therefore, it is important to warm the catalyst as early as possible. For the purpose, the manifold converter has been developed by using a metal (75Fe-20Cr-5Al) to support the catalyst [7.17]. The metal-supported catalyst has low ventilation resistance owing to large opening area compared with the ceramic-supported catalyst under the same number of cells, and has low heat resistance due to its welding structures with a small number of parts.

(2) Diesel-powered vehicles

The combustion mechanism of diesel engines is more complex compared with that of gasoline engines. At the last period of combustion process in the engine, diffusion-like combustion takes place and results in the formation of soot, and the residence time of burned gases at high temperatures is too short to oxidize the soot completely. Therefore, the soot (particulate) is to be emitted from diesel engines as well as NOx. Although fuels for diesel-powered vehicles contain a small quantities of sulfur, its emission can be neglected, and emissions of CO and HC are also of small amount due to large air-fuel ratio combustion. The technologies for reducing NOx emission from diesel engines tend to enhance the emission of soot. The formation mechanism of soot has not yet been well understood. The exhaust gas temperature is too low to use the catalytic converter for NOx reduction. Up to the present, adequate technologies for reducing simultaneously NOx and soot emissions have not yet been developed.

In order to reduce NOx emission at present, the following counterplans are used : (1) improvement of engine, (2) improvement of fuel injection system and (3) exhaust gas recirculation.

## 7.3 Formation and Control of SOx

Sulfur oxides ($SOx = SO_2 + SO_3$) are generated by oxidation of sulfur in fuels through combustion processes. At high combustion temperatures, $SO_2$ is generated in combustors and then emitted to the atmosphere, while at low temperatures, a part of $SO_2$ is oxidized to $SO_3$ in combustors. As mentioned before, good compliance with the environmental quality standards for $SO_2$ is attained by installing the fuel and flue gas desulfurization facilities. The flue gas desulfurization is mainly a wet-type and limestone ($CaCO_3$) is generally used, although $Ca(OH)_2$, $NH_4OH$, $Mg(OH)_3$, NaOH etc. are also used.

Figure 7.21 shows the flue gas desulfurization system developed by Ishikawajima-Harima Heavy Industries Co., Ltd. [7.18]. The exhausted gases from a boiler were introduced into the prescrubber to remove particulates, and after that into the absorber, where the $SO_2$ was absorbed by the spray of $CaCO_3$-water slurry. The desulfurization reactions are given by

$$CaCO_3 + SO_2 + 0.5\ H_2O \rightarrow CaSO_3\ 0.5\ H_2O + CO_2 \qquad (7.9)$$

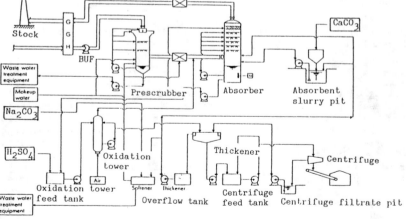

FIGURE 7.21 Flue Gas Desulfurization System [7.18].

A part of calcium sulfite ($CaSO_3 \cdot 0.5\ H_2O$) is oxidized by oxygen in the exhaust gas as

$$CaSO_3 \cdot 0.5\ H_2O + 0.5\ O_2 + 1.5\ H_2O \longrightarrow CaSO_4 \cdot 2H_2O \quad (7.10)$$

The remainder is oxidized to plaster ($CaSO_4 \cdot 2H_2O$) by air blown up from the bottom of the absorption tower and the remained $CaCO_3$ is also converted to plaster by the reaction with sulfuric acid.

$$CaSO_3 \cdot 0.5\ H_2O + 0.5\ O_2 + 1.5\ H_2O \longrightarrow CaSO_4 \cdot 2H_2O$$
$$CaSO_3 + H_2SO_4 + H_2O \longrightarrow CaSO_4 \cdot 2H_2O + CO_2 \quad (7.11)$$

The performance test of the desulfurization with 350 ppm inlet $SO_2$ and the gas flow rate of 168300 $Nm^3/h$ as for a 500 MW coal fired power station showed $SO_2$ removal efficiency of 96.8 % with the plaster of 99.1 % in purity and of 9.2 % water as by-products.

## 7.4 Simultaneous Removal of NOx and SOx by Electron Irradiation [7.19]

In order to remove simultaneously NOx and SOx from flue gases, a dry-scrubbing technology by using electron irradiation has been developed by the Japan Atomic Energy Institute and Ebara Corporation since 1971. Figure 7.22 shows the schematic diagram of denitrification and desulfurization by electron irradiation. After removing particulates in flue gases, the flue gases with the temperature of about 150°C are cooled to the temperatures in the range of 70°C to 90°C for optimum denitrification and desulfurization. After that, ammonia determined by the concentrations of NOx and SOx is added and electron is irradiated. The

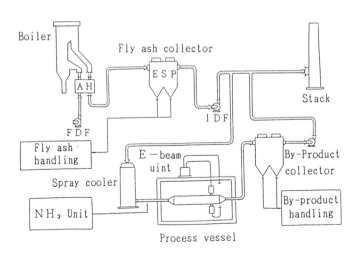

FIGURE 7.22 Schematic Diagram of Denitrification and Desulfurization by Electron Irradiation [7.19].

electron irradiation to the flue gases produces active species such as H, OH and $HO_2$ from $H_2O$, $O_2$ etc., which react with NOx and SOx to form ammonium sulfate-nitrate $((NH_4)_2SO_4 \cdot 2NH_4NO_3)$ and ammonium sulfate $((NH_4)_2SO_4)$ in the presence of ammonia $(NH_3)$. These mixed salts are recovered as a dry powder.

Figures 7.23 and 7.24 show the removal efficiencies of NOx and SOx versus the electron energy irradiated, respectively. Both removal efficiencies increase with increase in the electron energy. The removal efficiency of NOx is higher at higher vessel outlet temperatures,

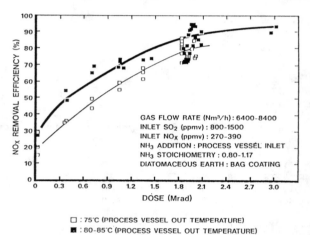

FIGURE 7.23   Removal Efficiency of NOx by Electron Energy [7.19].

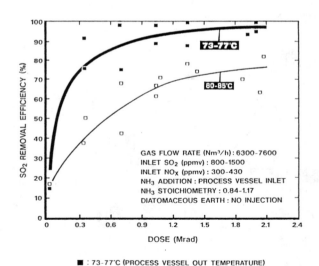

FIGURE 7.24   Removal Efficiency of SOx by Electron Energy.

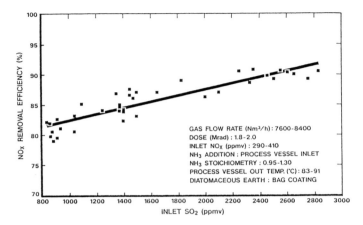

FIGURE 7.25  Removal Efficiency of NOx versus Inlet SOx Concentration.

while that of SOx is higher at lower vessel outlet temperatures. Figure 7.25 shows the removal efficiency of NOx versus inlet SOx concentration, and increase in the SOx concentration increases the removal efficiency of NOx.

## 7.5 Acid Rain

In the second half of 1970, damage to forests and agricultural products by acid rains was reported in Europe and North America. Recently, in Japan, such damage to forests, presumably, caused by acid rains has been reported. Since 1983, Environmental Agency has investigated the state of acid rain in Japan and found to have annual average values of about PH=4. Further, in 1988, 23 national ambient monitoring stations for acid rain were established [7.1].

## 7.6 Concluding Remarks

As mentioned above, in Japan, the concentrations of sulfur oxide (SOx) in 1986 could attain to the level lower than the value of the environmental quality standard at 98.9 % of air pollution monitoring stations. Otherhand, the concentrations of nitrogen oxide (NOx) have again increased since 1985. Recently, however, the emission control technologies for NOx have been successfully developed except for diesel-powered vehicles. For the latter problem of diesel engines, further development of technologies is urgently required for simultaneous reduction of NOx and soot emissions.


REFERENCES

[7.1]  Quality of the Environment in Japan, (1988), Environment Agency, Government of Japan.
[7.2]  Johnson, G.M. and Smith, M.Y., Emissions of Nitrogen Dioxide from a Large Gas-Turbine Power Station. Combust. Sci. and Technol., Vol.19, pp.67-70, 1978.

[7.3] Traynor, G.W., Girman, J.R., Apte, M.G., Dillworth, J.F. and White, P.D., J. Air Pollution Control Assoc., Vol.35, pp.231-237, 1985.
[7.4] Sano, T., $NO_2$ Formation in the Mixing Region of Hot Burned Gas with Cool Air, Combust. Sci. Technol., Vol.38, pp.129-144, 1984.
[7.5] Sano, T., $NO_2$ Formation in the Mixing Region of Hot Burned Gas with Cool Air - Effect of Surrounding Air, ibid, Vol.43, pp.259-269, 1985.
[7.6] Ishikawajima-Harima Heavy Industries Co.,Ltd.
[7.7] Formation Mechanisms and Controls of Pollutants in Combustion System, Ed. by The Japan Society of Mechanical Engineers, 1980 (in Japanese).
[7.8] Okazaki, K., Sugiyama, K. and Yuri, I., Study on the Formation Mechanism of Thermal NOx in Pulverized Coal Combustion, Trans. JASME, Vol.56-530(B), pp.3134-3141, 1990 (in Japanese).
[7.9] Masai, T., Morita, S., Akiyama, I. and Ohtsuka, K., Low NOx Combustion Technology for Pulverized Coal Fuel, Hitachi Review, Vol.34, pp.207-212, 1985.
[7.10] Aoyama, K. and Mandai, S., Development of a Dry Low Nox Combustor for a 120 MW Gas Turbine, Trans. ASME, J. Eng. Gas Turbines and Power, Vol.106, pp.795-800, 1984.
[7.11] Takeyama, K., Nakahara, T. and Mandai, S., Theoretical and Experimental consideration on NOx Formation Mechanism and Its Reduction, 11th Int. Congress on Combustion Engines (CIMAC), Vol.3, pp.207-230, 1975.
[7.12] Niigata Engineering Co.,Ltd.
[7.13] Ohasi, M., Development of Automotive Emission Control Catalysts, Catalyst,Vol.29, pp.98 , 1987 (in Japanese).
[7.14] Kawai, M. & Watanabe, H., Electronic Control for Fuel Injection System, J. Soc. Automotive Engineers Japan, Vol.32, pp.122-127, 1978 (in Japanese).
[7.15] Toyota Motor Corporation.
[7.16] Matsumoto, S., Miyoshi, N., Kimura, M. and Ozawa, M., Development of Thermal Resistant Catalysts, TOYOTA Engineering, Vol.38, pp.17-25, 1988 (in Japanese).
[7.17] Asano, M., Yoshimura, K. and Yasukawa, M., Development of Metal Supported Catalytic Manifold Converter, TOYOTA Engineering, Vol.40, pp.95-101, 1990 (in Japanese).
[7.18] Nakahira, T., Sakamoto, Y. and Kawamura, T., Operating Results of Flue Gas Desulfurization System for Unit Nos. 1 and 2 (500 MW Each) of Shin-Onoda Power Station of The Chugoku Electric Power Co., Inc., Ishikawajima-Harima Heavy Industries Co.,Ltd., Technical Report, Vol.27, pp.1-6, 1987 (in Japanese).
[7.19] Kawamura, K., Hirano, S., Hirano, Y., Maezawa, A., Aoki, S., Kaneko, M. Suzuki, R., Development of Electron Beam Dry Flue Gas Treatment Process (EBA Process) - Process Demonstration Operation in U.S.A. Coal fired Power Station, Ebara Engineering Review, Vol.141, pp.2-10, 1988.

# 8. RESEARCH AND DEVELOPMENT OF ALTERNATIVE ENERGY

## 8.1 PRESENT STATUS OF DEVELOPMENT OF ALTERNATIVE ENERGY TECHNOLOGY FROM ENVIRONMENT PROTECTION POINT OF VIEW

**Tadayoshi Tanaka**
Electrotechnical Laboratory
Agency of Industrial Science and Technology
Ministry of International Trade and Industry
Tsukuba Science City, Ibaraki 305, Japan

Japan lacks fossil fuel resources. Consequently, almost all fossil fuels are imported from abroad. Terefore, change in international affaires affects on Japan's politics and social life, as learned from experience of economic confusion caused by the oil crisis of 1973. For this reason, research and development(R & D) of alternavite energy technologies was initiated in July 1974, which was promoted as one of national energy development programs called Sunshine Project. Presently, their technical developments are being continued to put practical use under this project. However, Japan's dependency of primary energy resources on oil is still high among major advanced countries and energy supply structure is significantly weak. Futhermore, from indetermination of the recent political condition in the Middle East, the importance of security against supply and demand of petroleum in middle and long term is generally recognized with the increasing cost of oil.

On the other hand, pollution of earth's atmosphere is rapidly emerging as an important public policy issue. In these circumstances, development of alternative energy comes to be one of supremely important policy subjects for Japan to insure steadily dependable energy supply sources as well as protection of grobal environment. For this purpose, the second step of R & D of alternative energy technology is being promoted to resolve their problems under Sunshine Project. Presently, this project consists of the following four technical fields except nuclear power:

(1) Solar Energy
    a. Development of commercialization technology for solar photovoltaic power generation
    b. Development of solar energy systems for industrial use

(2) Geothermal Energy

       a. Development of geothermal prospecting and extraction technology
       b. Development of electric power generation systems using geothermal hot water
       c. Development of hot dry rock power generation system

(3) **Coal Liquefaction and Gasification**
       a. Development of coal liquefaction technology
       b. Development of coal-based hydrogen production technology
       c. Development of technology for coal gasification combined cycle power generation

(4) **Hydrogen Energy**
       a. Researches on solid-polymer-electrolyte water electrolysis and high-temperature water-vapor electrolysis

In addition, basic research on such subjects as ocean energy and wind power are also promoted under the Sunshine Project.

A part of major items in the Sunshine Project is now in the stage of developing large-scale test plants. Among above technologies, high priorities are given to the promotion of the following two projects for protecting environment and for establishing stable energy supply structure without affecting world oil supply-demand situation,

(1) Solar Energy
(2) Geothermal Energy

The reason why high priolities are given to these two technical developments is recognized by the following study. This analysis were carried out for the energy technology evaluation taking into account the emission of $SO_x$, $NO_x$ and $CO_2$ as well as the viewpoint of the energy security in Japan[8.1.1]. The method of the analysis and the system model structure of this study is the Market Allocation Model developed as one of the tasks(ETSAP) in the IEA. Typical example of major results of evaluation of alternative energy is shown in **Fig.8.1.1**. This result indicates the presumed amount of power generation during 40 yeares from 1985 to 2025, and is obtained by assuming that the currency in 1985 is adopted as a basis of the standard price and that the rate of the GDP reduces from 2.7%/y to 1.0%/y(average 1.6%/y). The unit of the vertical axis, EJ is exa-joules/year($10^{18}$J/year). Top column of the figure describes cost function and Min[P] expresses the result of the case minimizing only the cost. Coefficient of $CO_2$ means a fictitious penalty of $CO_2$ emission. Its unit is $/ton $CO_2$. In all the power generation systems, electric power supply depends on atomic power and gas with the rise of emission constraint of $CO_2$. In the field of alternative energies except for hydro-power, it is estimated that electric power is supplied by solar energy and geothermal energy as shown in the figure. Therefore, we think solar energy and geothermal energy will become future promising energy resources in Japan.

From above reasons, solar energy and geothermal energy not only play important roles in contributing to improvement of the future energy structure, but also have great potentiality of commercialization. Therefore, in this section, we described present status and future prospects of R & D of thier technologies in Japan as below.

**REFERENCE**

[8.1.1.] Koyama.S. and Ihara.S., Effect of Emission Constraint on the Structure of Energy System, ETL Tech. Report., pp80,1989 (Japanese)

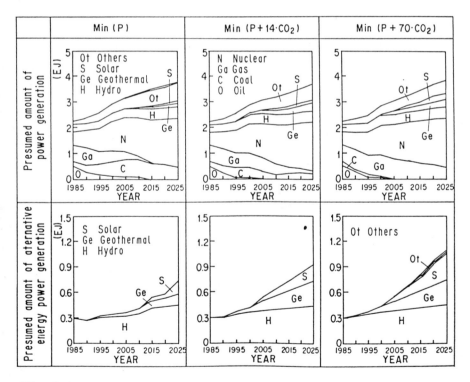

Fig.8.1.1 Future Formation of Electric Power Systems in Japan

# 8.2 CURRENT STATUS AND FUTURE PROSPECTS OF SOLAR ENERGY DEVELOPMENT

**Tadayoshi Tanaka**
Electrotechnical Laboratory
Agency of Industrial Science and Technology
Ministry of International Trade and Industry
Tsukuba Science City, Ibaraki 305, Japan

Solar energy, because of its inexhaustibility and cleanliness, is an ideal energy source and has been recognized as one of mankind's major energy options. In conversion of solar radiation into other forms of energy, heat energy use and the production of electricity have been of highest priority. Research and development of solar energy is being promoted in the following two technical fields of photovoltaic conversion and solar thermal energy utilization.

## 8.2.1 DEVELOPMENT OF PHOTOVOLTAIC POWER GENERATION SYSTEM

Photovoltaic (PV) power generation is one of the methods conversing directly solar energy into electricity by solar cell. In order to promote the development of practical applications of PV power generation system, the following subject areas are being addressed.

(1) Solar Cell Production Technology
(2) PV Power Generation System Technology

The most important objective of these programs is to develop the technology for manufacturing low cost-high efficiency solar cell and the system technology applied to various uses.

### [1] SOLAR CELL PRODUCTION TECHNOLOGY

It is known well that solar cells are used as a small electric power source in many fields such as pocket calculators, watches and telecommunication. As illustrated in **Fig 8.2.1**[8.2.1], the production of solar cell in Japan is rising markedly every year and accounts for a large portion of the production of solar cells in the world. For spreading use of solar cell, the cost milestone of solar cell is set up ¥500/Wp around 1990 and ¥100-¥200/Wp in the beginning of the 2000's. To attain this goal, the following three items are provided for technology of manufacturing low cost solar cell,

(1) Cristalline Solar Cell
   a. Material manufacturing
   b. Cast and sheet substrate fablication
   c. Junction and cell fablication
(2) Advanced Solar Cell
   a. Thin film substrate solar cell
   b. Tandem type solar cell
(3) Amorphous Solar Cell
   a. Low cost monosilane manufacturing
   b. Cell fablication

Crystalline sillicon cells are manufactured through each process of material product, substrate fablication and cell fabrication. Presently, technology of each manufacturing process excluding panel assembly reaches the practical level because the disired objectives shown in **Table 8.2.1** were attained. Technical development of crystalline solar cell finished in 1988.

After then, advanced solar cells of thin film substrate and tandem type, having a potencial of much higher efficiency and much lower cost, are developing to attain the objective shown in **Table 8.2.2**. As shown in this table, the size of high purity silicon thin film is below 200 μm thickness and 10 cm square. Conversion efficiency of cell made from it is estimated at above 18 %. On the other hand, the tandem type solar cell is made from amorphous-polycrystal silicon or amorphous-compound semiconductor, and the conversion efficiency of each cell is 14% or over with 10 cm square and 13% or over with with 30 cm x 40 cm rectangle.

It is predicted that the manufacturing cost of amorphous solar cell reduces markedly because it can be made from only a small amount silicon with mass and continuous production. Presently, their manufacturing technologies are promoted to obtain higher quality, larger area and higher reliability. Target of amorphous solar cell manufacturing technology is set as shown in **Table 8.2.3**.

From above matter, conversion efficiency of small area solar cell is increasing as shown in **Fig.8.2.2**. In early stage of development of silicon solar cell, main target of its technology is to reduce cost rather than to rise cell efficiency. Hence, the value of efficiency is changing maintaining the nearly same level during about twenty years from 1960 to 1980. After then, it increases markedly by the effort to obtain break-through technology for increase in efficiency. Presently, silicon solar cells are used in many PV power systems. On the other hand, conversion efficiency of amorphous solar cell increases linearly by improvement of manufacturing technology.

## [2] PHOTOVOLTAIC POWER GENERATION SYSTEM TECHNOLOGY

For the wide spread of PV power generation systems, it is also necessary to establish the technologies to assemble PV power systems and to transmit generated electricity to a power tramsmission line, altough it is necessary to reduce the costs not only of solar cell but also of peripheral devices such as storage battery and inverter. To attain this purpose, the following technologies are promoted with system development for various purpose.

(1) Cost Reduction of Peripheral Devices
    a. Module supporting frames
    b. Storage battery
    c. Inverter
(2) Improvement of Peripheral Technology
    a. On-site power trasmission line connection control technology

On the other hand, demonstration plants of PV power systems were already constructed at many sites to clarify possibility of introduction into the following sectors:

(1) PV Application Systems
    a. Single-family house
    b. Apartment house
    c. School building
    d. Car battery factory
(2) PV Power Generation Systems
    a. Distributed arrangement of panels
    b. Concentrated arrangement of panels
(3) PV/Thermal systems
    a. Concentrated type
    b. Flat plate type

The outlines of above systems are described in **Table 8.2.4**. Almost all solar cell panels used in these systems are composed of crystalline silicon solar cell. Among above systems, the operating results carried out in the concentrated arrangement PV power generation system are shown as typical example in **Table 8.2.5**[8.2.2]. System efficiency of this system is very low in comparison with existing power generation systems by oil or coal combustion, though conversion efficiency of small area solar cell is high as shown in Fig.8.2.2. If the construction cost of PV power systems would reduce, we feel that large scale PV power system is not suitable for Japan's limited terrain.

Wide spread of PV conversion technology does not progress presently in spite of above efforts. It is, therefore, expected that small PV power systems are used as one of the power supply systems in isolated islands or are operated in interconnection with existing power transmission lines.

## 8.2.2 DEVELOPMENT OF SOLAR THERMAL UTILIZATION TECHNOLOGY

Solar thermal utilization system consists of a collector obtaining effectively solar energy, a heat storage unit storing heat during a rainy day or at night when solar energy is not available, a heat transfer subsystem and an utilities for using effectively its heat. In the case where low temperature heat is collected from the sun, its heat is used for heating, cooling and hot water supply (Solar house). In the case where high temperature heat is collected, its heat is used for power generation (Solar thermal electric power system). The development of above solar thermal energy utilization technology were promoted with the development of materials and components.

### [1] LOW TEMPERATURE SOLAR THERMAL ENERGY UTILIZATION TECHNOLOGY

For effective use of low temperature solar thermal energy, technical development of air conditioning and hot water supply by solar energy was promoted with the aim of introduction to residential use, The following items were provided for this development.

(1) Solar Heating-Cooling and Hot Water Supply Systems
    a. Existing single-family house
    b. Newly built single-family house
    c. Multi-family house
    d. School building
(2) Development of Materials and Compornent
    a. Selective surface
    b. Solar Collectors

After above solar houses were constructed, about two years operation was continued. At the same time, the development of materials and components was promoted, and high-efficiency flat plate collectors and evacuated tubular collectors with selective surface(i.e., black crome) and solar driven absorption type refrigerator(LiBr solution) were developed for being used in above solar houses. These technologies are presently at the stage of spread or practical application. Number of solar systems installed in last six years is indicated in **Fig.8.2.3**[8.2.3]. As shown in this figure, number of solar system is decreasing every year. Futhermore, almost all solar systems are of hot water supply use and solar systems with air conditioning is few. We feel that solar systems do not spread exceedingly because it is expensive in comparison with an equipment using city gas or electricity even if it is used for hot water supply.

On the other hand, based on above technology, development of solar systems for industrial use is promoted to spread widely the use of solar thermal energy as well as residential use, because vast amount of energy is consumed in industrial sectors in Japan. The following industrial solar systems were constructed and were being operated.

(1) Fixed Heat Process Type Solar Systems
   a. Lumber drying system
   b. Warehouse for storing green vegetables

(2) Advanced Heat Process Type Solar System
   a. Refrigerating storehouse

Fixed heat process type solar system is developed with the aim of application of precess needing constant temperature heat. Above two demonstration systems were constructed. Lumber drying system developed is provided for commercial use. On the other hand, advanced heat process type solar systems are the one providing effectively either low or high temperature heat obtained from the sun. For this project, the first constructed solar system was applied dying process of a knitwear industry. After then, development of refgerating storehouse system was constructed. Temperature in a storehouse is kept within $-5\sim0$ C by an absorption type refrigerator(Refrigerant Trifluoroethanol, Absorbent N-methyl 2-pyrolidone) driven by the heat of about 140 C carried from a collector loop composed of evacuated tubular collectors. Other systems except lumber drying system do not attain to the stage of commercial use. This is the reason why system cost is very high. Especially, development of cost reduction of components such as collectors is progressed.

## [2] HIGH TEMPERATURE SOLAR THERMAL ENERGY UTILIZATION TECHNOLOGY

Electric power is one of the most important energies for progressing economical and public society. As mentioned above, Japan lacks fossil fuel energy. So, it is important subject to generate electricity by energy resources without dependence on abroad. Therefore, it is one of the effective means to generate electricity by high temperature solar energy. From above reason, as technical development of high temperature solar energy utilization systems, development of solar thermal electric power system(STEPS) and solar total energy system(STES), which generates electric power and heat energy for industrial and domestic uses, were promoted. Two types of STEPS constructed were central reciever type(CRS) and hybrid mirrors type(HMP). A solar total energy experimental facility(STEEF) was constructed for basic research of the STES. Configurations of the STEPS anf the STEEF was shown in **Fig.8.2.4** and their major specifications were described in **Table 8.2.6**[8.2.4]. A central reciever type system is general type of the STEPS constructed in many sites. A hybid mirrors type system is only one type in the STEPS constructed in the world. In this system, sunlight is concentrated by a number of plane mirrors and cylindrical parabolic reflectors as shown in Fig.8.2.4(c). The STEEF is constructed as shown in Fig.8.2.4(d) and is operated in three different oprating modes. According to the operating mode, the heat stored in the heat storage units is used for power generation and/or heat energy supply as process steam, air conditioning, and hot water.

Typical example of results obtained from the operation of the STEPS and the STEEF are shown in **Fig.8.2.5**. System availability,SA is defined as follows:

$$SA = \frac{\text{total amount of electric power or the sum of it and heat energy supply}}{\text{rated output power} \times \text{number of days in a month} \times 24 \text{ hours}}$$

The system availability for power generation of the hybrid mirror, $SA_h$, is the lowest due to higher collection temperature. On the other hand, system availability for the STEEF,$SA_e$, is nearly equal to that for the CRS,$SA_c$, although turbine efficiency of the STEEF is lower than that of the CRS. In the STEEF, when additional high and low temperature heat supply is included, for $SA_{e+ht+lt}$ is also indicated on this figure. In this case, there is only minor heat loss accompaied with conversion of heat energy into electricity. Accordingly, the values of $SA_{e+ht+lt}$ becomes large. From such a result, it is concluded that system availavility is improved by providing multiple operating modes such as conducted at the STEEF. Futhermore, in the case of utilizing solar energy in Japan, we believe from this results that it is very important to collect low and high temperature heat from the sun and to use it for multipurpose utilization such as power generation and heat energy supply.

In spite of above effots, the development of high temperature solar thermal utilization systems such as the STEPS is being discontinued because it is difficult effectively to obtain high temperature heat from the sun under poor solar condition in Japan. we can not find out till now the break-through utilization technology of high temperature solar thermal energy in Japan. However, the STEPS in the U.S.A. comes to be the stage of one of the commercial power generation systems. It is the SEGS plant(Solar Electric Generating System). This plant is the STEPS with the greatest capacity in the world as shown in **Table.8.2.7**[8.2.5], which is one of the distributed type STEPS collecting solar energy by cylindrical parabolic trough mirrors collectors. Though the outlines of SEGS I-VII is described in Table 8.2.7, we heard that the SEGS VIII with capacity of 80MWe was completed in 1989. Present total capacity of power generation of the SEGS is about 270 MWe.

We estimated the effect of the STEPS on $CO_2$ emission reduction, based on the results of the SEGS plant. According to the results of Table 8.2.7, the average efficiency of the SEGS plant is estimated about 34%. Total amount of the necessary heat energy to generate electric power of 270 MWe is about $6.83 \times 10^8$ Kcal/h, and is supplied from solar energy and natural gas boiler. According to another informations on the SEGS plant, the rate of heat energy supplied by solar energy is 53%. Assuming that $CO_2$ emission of natural gas is 0.022g/Kcal, the following value of $CO_2$ emission is restrained by utilizing solar energy,

$$6.83 \times 10^8 \times 0.022 \times 10^{-6} \times 0.53 = 7.96 \text{ tons/h}$$

This value is equivalent to carbon weight of 2.17 tons/h. Assuming that the operating period per day of the SEGS plant is 8 hours, amount of carbon emission reduction in a year is,

$$2.17 \times 8 \times 365 = 6,336 \text{ tons/year}$$

On the other hand, it is reported that about 54% of 5.1 billion tons/year(carbon equivalent) is remained in atmosphere. Assuming that this amount is generated by fossil fuel power generation systems and that their systems are replaced by the STEPS such as the SEGS plants, necessary area for construcing the STEPS is calculated as follows, based on total area of the SEGS I-VIII, about 1.69 $Km^2$,

$$1.69 \times 275,400/0.6336 = 73.5 \times 10^4 \text{ } Km^2$$

It is reported that land of about $6 \times 10^4$ $Km^2$ becomes desert every year and that the area of about $3,500 \times 10^4$ $Km^2$ is desert. Futhermore, the desert is the most suitable construction sites of the STEPS because it locates in good sunny condition. Therefore, we do not feel that above size of solar field for constructing the STEPS is wide. Therefore, we believe that the STEPS contributes to protect grobal environment and to save consumption of fossil fuel if it constructed in the desert region.

### 8.2.3 FUTURE PROSPECTS OF SOLAR ENERGY DEVELOPMENT

After the first oil crisis in 1973, Japan lacking fossil energy resources is promoting development of alternative energy technology. As described in this section, solar energy is one of the promising energy resources, although Japan has poor solar condition. Development of PV power systems and solar thermal utilization systems is progressed for commercial use. In spite of such an effort, the use of these systems does not spread widely. Therefore, the most improtant subject in these developments is to reduce the cost. To attain this purpose, R & D of solar energy technology would be continued in future.

[REFERENCES]

[8.2.1] Sunshine Project-Solar Energy Utilization Technology edited by NEDO, Solar Energy Div., P8,1989
[8.2.2] 10th. Annual Repot of NEDO Activities, PP109, 1990 (Japanese)
[8.2.3] Bul. of Solar System Development Assosiation, Vol.13, No.53, PP20-21, 1990 (Japanese)
[8.2.4] Tanaka.T., Tras. of the ASME. J. of Solar Energy Eng., Vol.111, PP318, 1989
[8.2.5] Kaarney.D., etc., Proc. of the 10th Annual ASME Solar Energy Conf., pp254, 1988

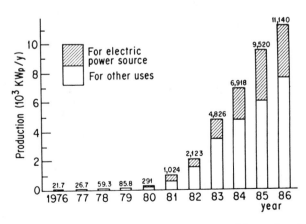

Fig.8.2.1 Yearly Solar Cell Production in Japan

**Table 8.2.1** Achieved Target of Crystalline Silicon Cell Manufacturing Technical Development

| Processes | Development Items | Achieved Target |
|---|---|---|
| Material | Low-cost Silicon manufacturing | ¥10/gram (100 ton/y) <br> ¥4.9/gram (1,000 ton/y) |
| Substrate | Cast substrate fablication | P type 38.6 Ω·cm <br> Life time 35 $\mu$sec <br> Cell conversion efficiency 15.8 % |
|  | Sheet substrate fablication | P type 50.0 Ω·cm <br> Life time 6 $\mu$sec <br> Cell conversion efficiency 12.3 % |
| Cell fablication |  | Maximum conversion efficiency 20.5 % (2cm square) <br> Cell conversion efficiency 15.8 % |

**Table 8.2.2** Target of Advanced Solar Cell Development

| Solar Cell | Items | Target |
|---|---|---|
| Thin film | High purity substrate | Impurity concentration <br> oxgen $2 \times 10^{17}$ atom/cm$^3$ or less <br> Carbon $5 \times 10^{16}$ atom/cm$^3$ or less <br> Minority carrier diffusion length <br> 150 $\mu$m or over <br> Ingot size 20cm x 20cm <br> Substrate thickness <br> 200 $\mu$m or less <br> Cell conversion efficiency 18% or over <br> (10 cm square) |
|  | High speed substrate manufacturing | Substrate size <br> 20cm x 20cm x 0.1mm <br> Production speed <br> 1 wafer/min or over <br> Cell conversion efficiency (10cm square) <br> 14 % or over |
|  | New cell design and fabrication | Cell conversion efficiency (10 cm square) <br> 18 % or over <br> (15 cm square) <br> 17 % or over <br> (2 cm square) <br> 23 % or over |
| Tandem type | Amorphous-polycrystal silicon | Cell conversion efficiency (10 cm square) <br> 14 % or over |
|  | Amorphous compound | Cell conversion efficiency (30 cm x 40cm) <br> 13 % or over |

Table 8.2.3 Target of Amorphous Solar Cell Manufacturing Technology

| Items | Target |
|---|---|
| Higher quality | Cell conversion efficiency 12% or over (10cm square)<br>Degradation rate after one year 10% or less |
| Larger area | Substrate size 30cm x 40cm<br>Cell conversion efficiency 10% or over<br>Degradation rate after one year 15% or less |
| High reliability | Initial cell conversion efficiency 11% or over (10cm square)<br>Degradation rate after one year 3% or less |

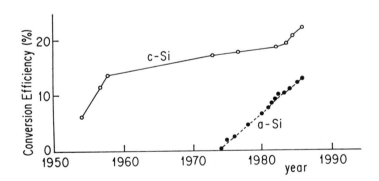

Fig.8.2.2 Conversion Efficency of Solar Cell
(c-Si Crystalline Silicon Solar cells,
a-Si Amorphous Solar Cells)

Table 8.2.4 Outline of PV Power Systems

| Systems | Location (Prefecture) | Solar Cell (kW) | Battery Capacity (kWh) |
|---|---|---|---|
| Single-family house | Yokosuka (Kanagawa) | 3 | 14 |
| Apartment house | Tenri (Nara) | 20 | 114 |
| School biuilding | Tsukuba (Ibaraki) | 200 | 576 |
| Car battery factory | Kosei (Shizuoka) | 100 | 500 |
| PV power systems | | | |
|    Distributed | Ichihara (Chiba) | 200 | 400 |
|    Concentrated | Saijo (Ehime) | 1,000 | 1,800 |
| PV/thermal | | | |
|    Concentrated | Saka-Machi (Hiroshima) | 5(ele.) 25(thm.) | 37 |
|    Flat plate | Hirazuka (Kanagawa) | 3.2(ele.) 21(thm.) | 38 |

Table 8.2.5 Operating Results of PV Power Generation System of Concentrated arrangement

| Items | 1986 | 1987 | 1988 | 1989 | Average |
|---|---|---|---|---|---|
| PV power generation (kWh/year) | 1,545 | 1,647 | 1,555 | 1,501 | 1,562 |
| Net power generation (kWh/year) | 1,389 | 1,544 | 1,460 | 1,416 | 1,452 |
| PV panel efficiency (%) | 7.6 | 7.8 | 7.6 | 7.5 | 7.6 |
| System efficiency (%) | 6.8 | 7.3 | 7.1 | 7.1 | 7.1 |

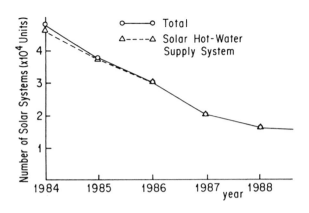

**Fig.8.2.3** Number of Solar Systems installed in Last Six Years

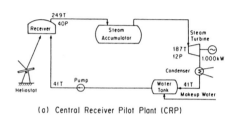

(a) Central Receiver Pilot Plant (CRP)

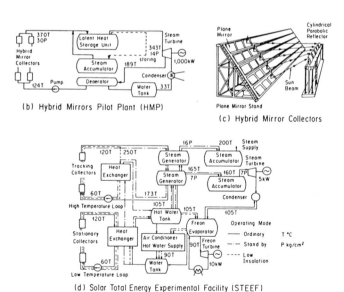

(b) Hybrid Mirrors Pilot Plant (HMP)

(c) Hybrid Mirror Collectors

(d) Solar Total Energy Experimental Facility (STEEF)

**Fig.8.2.4** System Configuration of Solar Thermal Electric Power Plants And Solar Total Experimental Facility

**Table 8.2.6** Typical Specifications of the STEPS and the STEEF

| Items | CRP | HMP | STEEF |
|---|---|---|---|
| Construction Site | NiO (Kagawa) | NiO (Kagawa) | Tsukuba (Ibaraki) |
| Land Area (m²) | 50.000 | 50.000 | 4.000 |
| Heat Collection Temperature (°C) | 249 | 380 | 250 120 |
| Collector Fluid | Water | Water | Organic Oil |
| Turbine Inlet Temperature (°C) | 187 | 343 | 165(Steam) 90(Freon) |
| Steam Condition for Power generation | Saturated | Super heated | Saturated |
| Turbine Efficiency (%) (at Rating) | 17 | 22 | 5(Steam) 5(Freon) |
| Rated Output | 1.000 kWe | 1.000 kWe | 15 kWe 45 kWt |
| Heat Storage Capacity | 1.000 kWe x 3h | 1.000 kWe x 3h | 15 kWe x 1h 45 kWt x 1h |
| Operating Period | Sep. 1981 ~Mar. 1984 | Oct. 1981 Mar. 1984 | Sep. 1982 ~Jul. 1985 |

\* except heat load of air conditioning and hot water supply (Total amount 80 kWt)
kWe : electric   kWt : thermal

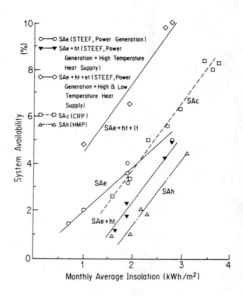

**Fig.8.2.5** System Availability of the STEPS and the STEEF

**Table 8.2.7** Basic Characteristics of the SEGS Plant

| Plant | First Full Operating Year | Status | Turbine Capacity (MWe net) | Solar Temp. (°C) | Field Size (m²) | Turbine Cycle Efficiency | | Annual Output (MWh net) |
|---|---|---|---|---|---|---|---|---|
| | | | | | | Solar | Boiler | |
| I | 1985 | Operational | 13.8 | 307 | 82960 | 31.5 | — | 30100 |
| II | 1986 | Operational | 30 | 315 | 165376 | 29.4 | 37.3 | 80500 |
| III | 1987 | Operational | 30 | 349 | 203980 | 30.6 | 37.4 | 85050 |
| IV | 1987 | Operational | 30 | 349 | 203980 | 30.6 | 37.4 | 85050 |
| V | 1988 | Operational | 30 | 349 | 233120 | 30.6 | 37.4 | 91820 |
| VI | 1989 | Construction | 30 | 390 | 188000 | 37.5 | 39.5 | 90575 |
| VII | 1989 | Construction | 30 | 390 | 183120 | 37.5 | 39.5 | 94410 |

# 8.3 CURRENT STATUS AND FUTURE PROSPECTS OF GEOTHERMAL ENERGY DEVELOPMENT

**Yukio Yamada**
Mechanical Engineering Laboratory
Agency of Industrial Science and Technology
Ministry of International Trade and Industry
Tsukuba Science City, Ibaraki 305, Japan

## 8.3.1 INTRODUCTION

It is almost 85 years since the advent of geothermal power generation when the first experimental geothermal power plant at Larderello in Italy generated 0.5 kW electricity in 1904. Subsequently, the industrial use of geothermal energy started in 1913 through the development of a 250 kWe power plant also in Italy. Since then the geothermal power generation has been widely developed in the world and the growth of installed plant capacity as of 1985 is shown in Fig. 8.3.1.[8.3.1] The average annual growth rate was 6% to 6.5% in the period from 1950 to 1970, 13.5% from 1970 to 1980, and 17.5% from 1980 to 1985. In 1988, the total capacity of over 6,000 MWe geothermal power plants has been installed all over the world as shown in Table 8.3.1.[8.3.2] Most of the power plants are operated with the annual working ratio of 0.75 to 0.95, which is significantly higher than any other power plants such as fossil fuel-fired or nuclear power plants, thus generating $3.7 \times 10^{10}$ kWh annually in 1988. But this amount is still less than 1% of the total annual electricity consumption in the world.

The direct use, or multi-purpose use, of geothermal energy as heat source has recorded total 4,500 MWt in 1984 excluding the use of balneology.[8.3.3] The direct use is also growing with the growth rate of 8.5% in the period from 1975 to 1984. This growth rate is not so high as that of the geothermal power plant because the cost of the direct use is comparatively higher than other conventional heat sources.

Although the geothermal power generation as well as the direct use of geothermal energy has not established its position as one of the major energy sources as described above, geothermal energy is becoming highly expected as one of the alternative energy resources because of its cleanliness in the recent upheaval of the environmental problem. The following describes the current status and future prospects of geothermal energy utilization.

## 8.3.2 TECHNOLOGY AND CURRENT STATUS OF GEOTHERMAL POWER GENERATION

The major geothermal resoources for power generation are steam and brine spouting out from the underground reservoirs through geothermal wells. In addition to reservoirs, hot dry rocks and magmas are also considered as the energy sources, and the technologies for heat extraction from these sources are being developed as the future technologies.[8.3.4] Geothermal wells drilled into reservoirs containing steam and brine can produce dry steam (super heated and containing no brine), brine, or mixture of steam and brine. Usually, wells producing brine only cannot be used for power generation because of the poor energy quality. The types of geothermal power plants are decided according to the fluid obtained from the wells as in the following list, and the exhaustive reviews of geothermal power generation are found in the literature.[8.3.5-6]

(1) Dry steam ---------------- Steam turbine power generation
(2) Steam and brine ------- Single-flash steam turbine power generation
------- Multi-flash steam turbine power generation
------- Binary cycle power generation
------- Total-flow expander power generation

Most of the geothermal power plants in the early time period used dry steam only, and even for the modern technology it is still true that the most economical geothermal power generation is achievable when only high temperature dry steam is obtained from geothermal wells. But, recent technological development has made it possible to utilize the energy of brine very effectively. Specifically, the average steam to brine flow ratio lies around 1:4 in Japan, and only a few wells produce dry steam only. Therefore, it is highly desired to utilize the energy of brine for effective use of geothermal energy.

## [1] STEAM TURBINE POWER GENERATION

Steam-turbine power generation is the easiest and most economical system, and technologically it is believed to be fully developed. Single-flash power generation, which separates brine from dry steam and uses dry steam only for power generation, forms a relatively simple plant, but it wastes all the energy of brine. Multi-flash power generation uses low pressure steam generated by flashing the separated brine, and has become more popular in the newly constructed geothermal power plants because of its potentiality of economical operation even for the wells producing only brine.

The United States has almost half of the world geothermal power plants as shown in Table 8.3.1. Most of the geothermal power plants in the United States are located in California. Especially, The Geysers area in central California generates about 2,000 MW, which is almost 70% of the United States total geothermal power generation and is equivalent to two large nuclear power plants. The quality and capacity of steam produced in The Geysers are superior to any other geothermal sites in the world and the capacity of the installed power plants is still growing.

Steam turbine power generation including the flashing type needs to eject non-condensable gases contained in geothermal steam and accumulated in condensers. These non-condensable gases such as $CO_2$ and $H_2S$ induce pollution problems at some geothermal sites. This problem must be considered particularly at the sites in national parks or close to residential areas.

## [2] THEORETICAL BACKGROUND OF BINARY CYCLE POWER GENERATION

Fundamental binary cycle power generation is depicted in Fig. 8.3.2. Heat of brine is transferred to another medium having a lower boiling temperature by a heat exchanger consisting of a preheater and an evaporator. And a turbine is driven by the gas of the working medium to generate power. The cycle of the medium is a so-called Rankine cycle. The irreversible process experienced by the heat exchanger (preheater and evaporator) produces the largest loss of available energy (or exergy) due to the temperature difference between the brine

and the working medium. In order to reduce this loss, modifications of the fundamental cycle have been devised and four of them are as follows.
(1) steam assisted cycle
(2) multi-stage evaporation cycle
(3) super-critical cycle
(4) down-hole heat exchanger/pump cycle.

The steam assisted cycle uses steam in addition to brine from the well head. The steam is used to evaporate the working fluid as shown in Fig. 8.3.3. The resultant temperature difference between the brine and the working medium becomes much smaller, and the irreversible loss is reduced. This cycle was tested by one of the prototype plants of "Sunshine Project" of Japan.[8.3.7]

In the multi-stage evaporation cycle shown in Fig. 8.3.4, the working medium is evaporated successively at several discrete pressures. The irreversible loss at each pressure stage is small and the total loss of the cycle becomes smaller than the fundamental cycle. Infinite stages can provide the minimum loss, but practically two to four stages are optimum from the economical view point because more complicated systems and higher costs are required for multi-stage cycles.[8.3.8]

The super-critical cycle pressurizes the working medium to a higher pressure than the critical pressure. Therefore, no phase change takes places during the heat exchange process, and the temperature difference between the brine and the medium can be kept small to reduce the loss. However, a greater pumping power required to pressurize the medium makes this cycle impractical.

The down-hole heat exchanger/pump cycle deliberately eliminates the medium pump of the super-critical cycle by use of the depth of a well. The schematics of this sophisticated system is depicted in Fig. 8.3.5 (a), and its thermodynamic cycle is shown in Fig. 8.3.5 (b). The condensate of the binary cycle medium is sent to a heat exchanger installed inside the well. During flowing down in the heat exchanger, the medium is heated by the counter-current brine flow and simultaneously pressurized by the head of the medium itself, and finally the medium pressure exceeds the critical pressure. At the bottom of the heat exchanger, a small turbine driven by the medium at a super critical pressure and temperature is equipped to drive the brine pump, i.e., a down-hole pump. The medium with lower density after the expansion through the small turbine flows up toward the power turbine on the ground to generate power, and the cycle closes. The down-hole pump pressurizes the brine so that the brine does not flash during rising the well. By adjusting the heat exchanger size, the temperature difference inside the heat exchanger can be controlled as shown in Fig. 8.3.5 (b), resulting in the minimum loss of available energy. This type of down-hole heat exchanger/pump system was devised and tested by Sperry Research Center in US as described later.[8.3.9]

These modifications of the fundamental binary cycle can increase the thermal efficiency, but at the same time increase the cost. Therefore, what type of the binary cycle should be used must be determined by considerations of various aspects at the particular geothermal site.

[3] CURRENT STATUS OF BINARY CYCLE POWER GENERATION

The binary cycle power generation utilizes brine by transferring its energy to other working media having lower boiling temperatures. The Rankine cycle of the other working medium drives a turbine to generate power. The binary cycle power generation is considered to be the most effective to utilize the energy of brine. The most popular binary cycle working media are chloro-fluorocarbons (CFC) such as R-114 or R-113, and alkane such as iso-butane or pentane. Two 1 MW test plants of binary cycle power generation were developed by the "Sunshine Project" of the Japanese government in 1978,[8.3.7] and the effort to develop 10 MW binary cycle power plants is being continued. China also constructed binary cycle power plants before, but currently the United States of America is the only one who commercially generates electricity by the binary cycle power generation systems.

One of the two binary cycle power plants developed by the Japanese government used R-114 as the working medium and generated maximum 1,024 kW electricity from 140°C geothermal brine. Figure 8.3.6 shows this binary cycle which is a typical simple Rankine cycle.[8.3.7] The other binary cycle power plant used iso-butane which was heated by 130°C geothermal steam and brine through the steam assisted cycle shown in Fig. 8.3.3. This cycle has an advantage that the temperature difference between the heating media and the working fluid is kept minimum throughout the heating mode. Thus the available energy loss caused by the temperature difference in the heat exchanger is kept minimum as stated in the previous section. The in-plant power requirement for the two test plants was as high as 40%, and the net thermal efficiency was between 9% and 10% for 1 MW gross output power. The major in-plant power requirement came from the working medium feed pumps and the fans for cooling towers. R&D to reduce the in-plant power requirement was conducted since then, and 20% of the in-plant power requirement has been believed to be reasonably achievable. But 20% is still much higher than 5% to 6% for the multi-flash steam power plants.

The United States currently has 144 MW binary cycle geothermal power plants for commercial use. Recently, a small sized 1 MW binary cycle power unit has been developed by Ormat Energy Systems Company, and 20 to 24 units have been installed in one place to generate 20 MW to 24 MW electricity at Imperial Valley in southern California. Each unit uses pentane as the working fluid. Magma Power Company have developed binary cycle power plants using iso-butane and pentane as the working medium and generating total 12 MW electricity also in southern California.[8.3.10]

The problems for the existing binary power plant technology to be widely used for commercial base are the high cost and the environmental impacts of CFCs. The present high cost will be resolved by R&D to reduce in-plant power requirement, in particular. In order to resolve the environmental problem caused by CFCs, alternative working media such as R-123 are considered and their fundamental physical properties will meet the requirements from the cycle performance and environmental points of view.

The binary cycle power generation will be more popular if a new technology for geothermal fluid production is successfully developed. The new technology is the down-hole pump briefly described in the above. The down-hole pumps extract geothermal fluids as brine as it stays in the reservoirs. The following describes R&D of the down-hole pumping system.

## [4] DOWN-HOLE PUMPING SYSTEM

When mixture of steam and brine is obtained from a geothermal well, it is usually the result of flashing of pressurized brine at the bottom of the well. The pressurized brine staying in the reservoirs flashes due to the decrease of head pressure during rising through the well, and the temperature of the geothermal fluid obtained at the wellhead becomes lower than that at the well bottom, resulting in the lower conversion efficiency of power generation. In order to keep the temperature and pressure of the geothermal fluid at the reservoirs, or in order to extract brine from non-spouting wells, the technology of down-hole pumping system is applied in some cases. This technology also has other advantageous aspects, such as to prevent scaling from deposition caused by flashing and release of $CO_2$, to prevent pollutant gases from being generated and ejected into the atmosphere, and to enhance the brine flow rate.

Several types of down-hole pumps have been devised as listed in the following.
    (1) shaft-driven down-hole pump
    (2) down-hole heat exchanger/pump.
    (3) submersible motor-driven down-hole pump

The shaft-driven down-hole pumping systems developed by Peerless Pump Company were tested in the United States, but all of them were believed to have stopped working because of a trouble of the bearings connecting 240m- long shafts. The down-hole heat exchanger/pumping system, which is very sophisticated and expected to provide a high conversion efficiency by use of a super-critical pressure binary cycle as stated before, was also built and tested in the United States. But its R&D was terminated after a test operation in an actual well in 1977.

New Energy and Industrial Technology Development Organization (NEDO) of Japan has been developing a submersible motor-driven down-hole pumping system since 1983. This program is one of "The Sunshine Project" for R&D of alternative energy sources and is called as the DHP program.[8.3.11] The schedule of the DHP program of Japan is shown in Table 8.3.2, and the goals as well as the specifications of the first, second pumps are shown in Table 8.3.3. The pilot pumping system (No.1 machine) was made and successfully tested at laboratory and in an actual geothermal well. The brine temperature ranged from 140°C to 170°C, and the driving power was 100 kW with 50 t/h brine flow rate and 300m pumping head. The prototype pumping system (No.2 machine) consisting of two 100kW pumps connected in series is now being developed. The brine temperature is supposed to range 170°C to 200°C, and the brine flow rate and pumping head are 100 t/h and 340m, respectively. The demonstration down-hole pumping system (No.3 machine) is planned to have a motor power of 500 kW and a brine flow rate of 300 t/h with a pumping head of 400m. Once this down-hole pumping system is developed, considerable geothermal energy will be available through the very effective use of geothermal fluid stored in the existing and future reservoirs.

Other technologies for heat extraction from non-spouting wells are those of very large heat pipes,[8.3.12] and down-hole coaxial heat exchangers.[8.3.13] Both technologies do not extract geothermal fluid but only heat from geothermal wells by exchanging heat between the brine in the wells and the working fluids in the heat pipes or the well heat exchangers. But since these technologies are totally dependent on the thermal conduction in the reservoirs, their heat extraction rates are limited to a very small range compared to the conventional technologies to extract geothermal fluids. Their application will be found in some special occasions such as in the areas without commercial power supplies, or in the case that only the direct use of heat is considered.

## [5] TOTAL-FLOW EXPANDER POWER GENERATION

Alternative method to generate power effectively using the energy of brine associated with steam is to operate total-flow (or two-phase-flow) expanders. High pressure steam and brine from the well head are directly introduced to the total-flow expanders which expand high pressure steam and brine simultaneously. If the performance of the total-flow expanders are good enough, thermal efficiencies of such systems will be higher than that of the flashing steam turbine plants or even that of the binary cycle power generation, because these total-flow expanders do not need heat exchangers and working medium pumps. Many types of total-flow expanders have been devised and tested. Major types are (1)brine impulse turbines, (2)screw expanders, (3)rotary-type expanders, (4)bi-phase turbines, and so on.

(1) The brine impulse turbine shown in Fig. 8.3.7 is driven by a two-phase mist flow produced by a specially designed nozzle at Lawrence Livermore Laboratory.[8.3.5, 8.3.14] The turbine efficiency was reported to have achieved about 40% with the average droplet diameter of 2.3 mm. Higher efficiency was supposed to be achieved with smaller droplet diameters and for the brine temperatures higher than 200°C. Currently this type of turbine is not used in the geothermal field, but in the field of waste heat recovery in industrial factories where higher temperature waste water is found.[8.3.15]

(2) The screw expander has two screw type rotors, and the gap between the two screws expands two-phase brine as they rotates. A helical screw expander modified from a Lysholm engine was tested also at Lawrence Livermore Laboratory in 1977, and was reported to have achieved the maximum 53% of the turbine efficiency.[8.3.16]

(3) A fundamental research of the rotary type expanders was conducted by modifying a commercially available rotary engine for passenger cars. The experimental results indicated the turbine efficiency as high as 60% at a certain condition.[8.3.5] NEDO in Japan developed another type of rotary expander which had the inner and outer rotors as shown in Fig. 8.3.8.[8.3.17] The shape of the inner rotor is a three-node epi-trochoid and the outer rotor envelopes the inner rotor. The rotation ratio of the inner rotor to the outer one is 4:3 to form a cycle consisting of intake, expansion, exhaust and partial compression in the volumes encompassed by the rotors and casing. A 300kW prototype expander was built and tested in 1982, and marked the maximum turbine efficiency of 54% at 1200 rpm engine speed.

(4) The bi-phase turbine consists of three major components; a two-phase nozzle, a rotary separator, and a liquid turbine.[8.3.18] As shown in Fig. 8.3.9, the rotary separator receives the kinetic energy of high speed water generated by the two-phase nozzle and the steam turbine blades attached to the rotary separator convert the kinetic energy of the high speed steam. The enthalpy of the steam flowing out the rotary separator is recovered by a conventional steam turbine, and the kinetic energy of water rotating together with the rotary separator is converted to power by a liquid impulse turbine. This bi-phase turbine is reported to generate 40% more power than the simple flashing steam turbine system. A

1.6 kW plant was tested at a geothermal site in Utah, U.S.A., in 1983, and the turbine efficiency was evaluated as 48%. This system was also successfully applied to waste heat recovery at a steel industry in Japan.

Generally speaking, it is supposed to be difficult for any type of the total-flow expander to achieve turbine efficiencies higher than 60% because of the atomization problem of brine, the sealing problem, and so on. However, the efficiency is satisfactory for steam/brine two-phase flow at temperatures higher than 250°C. And because of relatively lower expansion ratio in the total flow expanders, the exhaust two-phase flow usually carrys enthalpy high enough for further expansion by steam turbines or by binary cycle power generation. Therefore, the total-flow expander will often be used as a topping cycle machine for a combined power generation driven by high temperature geothermal fluid or waste water from high temperature industrial processes.

### 8.3.3 DIRECT USE OF GEOTHERMAL ENERGY

The wide variety of the direct use of geothermal energy is roughly classified to three categories as follows.[8.3.19]
  (1) Residential and Commercial Use
   a. Space heating and cooling
   b. Hot and cold water supply
   c. Waste treatment (disposable, bio-conversion, etc.)
   d. Refrigiration
   e. Deicing
   f. Road heating (snow melting, deicing)
  (2) Agriculture and Related Areas
   a. Crops (greenhouses, hydroponics, heated soil, etc.)
   b. Animal husbandry (heating and cleansing needs of cattle, swine, chickens, etc.)
   c. Aquatic farming (fish breeding, hatching, growing, etc.)
   d. Processing of agricultural products (waste disposal or conversion, drying, fermentation, canning, etc.)
  (3) Industrial Processes
   a. Chemical production (acids, fertilizer, fuel, etc.)
   b. Pulp treatment
   c. Mining (heat, water)
   d. Drying (cement, clay, fish, etc.)
   e. Water desalination / distillation
   f. Mineral recovery from hydro-thermal fluid
   g. Waste treatment and disposal

The temperature ranges for the various uses are schematically expressed by Fig. 8.3.10, which is known as the Lindal chart.[8.3.20] High temperature geothermal resources can be effectively used by cascading as their temperatures decrease. According to the United States national energy budget in 1968, $15.4 \times 10^{12}$ MJ of energy use can be satisfied by heat at temperatures less than 100°C, and this amount of energy is almost 25% of the total energy budget of the U.S., and presumably similar to the other countries in recent years. Therefore, a considerable portion of the total energy consumption can be supplied by geothermal energy resources at temperatures less than 100°C, if they are available where they are needed.

Table 8.3.4 shows the capacities of the direct use facilities, the annual total heat outputs, and the working ratios in different countries, and Table 8.3.5 shows them for different uses in Japan (excluding balneological use).[8.3.21] A significant part of the direct use in Iceland is for space heating and hot water supply. About 300 MW at temperature range between 60°C and 80°C is supplied. The high load factor of New Zealand results from their industrial use. In Japan, the applications in agriculture and aquaculture except balneology occupy a large part, and especially the load factor of fish breeding is extremely high.

Cascading use according to the temperature levels is well known by its capability of effective use of thermal energy. One of the typical example of the cascading use is found in a rural area, Akita prefecture of Japan as shown in Fig. 8.3.11.[8.3.21] Initial 75°C hot water is used for heating and hot water supply for individual houses and schools first, and subsequently 65°C water heats water for swimming pools. Finally water at 25°C is supplied to rice fields to raise the temperature of irrigation water, and supplied for road heating for snow melting. A new project of another direct use of geothermal energy is planned in Iwate prefecture.[8.3.22] This project utilizes waste brine from a geothermal power plant. In addition to the simple direct use of the waste brine, the project is planning to construct a binary cycle power generation in order to be economically viable.

In the direct use of the geothermal energy, it is very rare that the original brine is directly used as the working fluid throughout the system, because of its chemical characteristics and the disposal or reinjection problem. Usually surface water is heated by the heat exchange process with the brine, and is distributed. Therefore, securing of surface water is indispensable for the direct use. Also the areas requiring the direct use should not be far from the production wells to be economically competitive with other conventional energy sources. The cost of the distribution network is high because of low energy density of hot water compared to gas or electricity networks.

## 8.3.4 PROBLEMS AND FUTURE OF UTILIZATION OF GEOTHERMAL ENERGY

The geothermal power generation occupies only less than 1% of the total world power generation stated before, and this ratio is not supposed to vary to a great extent in the future. However, as shown in Fig. 8.3.1, the recent annual growth rate marks higher than 10%, indicating about 20,000 MWe by the year 2,000 which is more than three times the present value. From the global environmental point of view, geothermal power generation is expected to grow with a higher rate in the near future. But, the cleanliness alone, which is the advantage of the geothermal energy over the fossil-fuel fired and nuclear power plants, cannot be a motivation for achieving a significant growth of the geothermal power generation. There are three kinds of problems affecting the further development of geothermal energy use. These are (1)technical, (2)economical, (3)social and institutional problems.

(1) TECHNICAL PROBLEM

The size of the geothermal power plant ranges usually from 10 MW to 50 MW according to the scale merit consideration. However, as the development of the geothermal resource exploration continues, the number of smaller-sized power plants will increase. These medium and small-sized geothermal power plants are expected to develop in the near future. Various new ideas and techniques should be devised and developed for such smaller-sized plants to be economically successful. For example in Japan, a cogeneration system supplying electricity and hot water has been implemented at a hotel located in a hot spring resort area in Kyushu district. This system utilizes the steam which was wasted before since steam was unnecessary for balneology. That steam, in spite of its small capacity, generates 100 kW electricity by being introduced into a back-pressure type turbine which does not need a condenser. This system is economically viable and replaced significant part of commercial electricity which they bought before. Like this cogeneration system, power generation of 100 kW and below will be more popular and effective to utilize wasted steam or poor quality geothermal fluids which are not economically viable by the current technologies.

Development of the down-hole pumping system is another expected technology for geothermal energy. As stated before, its success will increase the geothermal resources and thermal efficiency of geothermal power plants, and will decrease the problems associated with scaling and deposits.

(2) ECONOMICAL PROBLEM

Table 8.3.6 roughly compares the estimated relative costs associated with the construction of 50 MW geothermal power plants of a single-flash steam turbine, a double-flash steam turbine, and a combined single-flash steam turbine and binary cycle. Among the facility costs, the drilling cost of geothermal wells occupies a greatest part of the geothermal fluid production facility. The percentage of successful drilling is currently about 60% for both production and reinjection wells. The increase of the successful drilling ratio by the improved exploration technique will reduce the facility costs. For the binary cycle power generation, if technology development can reduce the in-plant power requirement of current 20% to future 5% or 6% equivalent to the flashing steam turbine power plants, the cost of electricity will be dramatically reduced. Then the commercial implementation of the binary cycle plants will become much more easier than it is presently.

For the direct use of geothermal energy, economical competition with fossil fuels is the critical problem. The production and reinjection well costs occupies a greatest portion of the initial capital cost if the wells are drilled only for the direct use. Therefore, if the residual brine which used to be reinjected without use in the geothermal power plants is obtained, the well cost might be eliminated. The distribution cost for small users will also be significant, but will be reasonable for large users, particularly for users near population centers in developing areas.

(3) SOCIAL AND INSTITUTIONAL PROBLEMS

For small-sized geothermal power plants, flexibility is necessary in selecting a most appropriate power generation system. That system will be any one of or combination of the various power generation systems such as flashing steam turbines, back-pressure steam turbines, binary cycle systems, total-flow expanders, etc., according to the capacity and temperature of the available geothermal fluid and to the purpose of its use. Another important factor to be stressed is the connection with the district society through the combination of the power generation and direct use, and through the environmental impact to the site. Also, the connection with the commercial electricity network is indispensable for such small-sized geothermal power plants to cope with a sudden break down and to be economically successful. Therefore, from the political point of view, regulatory and institutional considerations such as tax incentive programs are necessary to aid the development of small-sized as well as large-scale geothermal energy use.

These three problems are closely inter-connected each other. Technical development will improve economics, and will lead to less environmental impact to the surroundings of the sites. Then the public acceptance (PA) will be obtained easily and might alleviate the regulations. Favorable institutional environment motivates the development of the overall geothermal energy use.

## 8.3.5 SUMMARY

The geothermal energy, the only one energy resource originating from the earth, occupies only a small portion of the total world energy demand. Today, it is increasingly expected to be developed because of its smallest impact to the environment. However, the geothermal energy has not developed as it has been expected so far because the difficulty of finding appropriate sites and the large initial capital cost have discouraged the motivation of development. In order to solve these problems, further technical development is highly necessary as well as the social and institutional aids to implement the developed technology. Politically the smooth connection and adjustment between the energy related communities are important, in particular. The authors express a sincere desire that the use of the geothermal energy will secure its established position to resolve the global energy and environmental problems in the future.

[REFERENCES]
[8.3.1] Cataldi, R., and Sommaruga, C., Background, Present State and Future Prospects of Geothermal Development, Geothermics, Vol.15, No.3, pp.359-383, 1986. (Japanese version, Geothermal Energy, Vol.12, No.3, pp.248-266, 1987).
[8.3.2] Tawara, M., Engineering of Geothermal Power Plants, Geothermal Energy, Vol.14, No.4, pp.317-336, 1989 (in Japanese).
[8.3.3] Gudmundsson, J.S., Direct Uses of Geothermal Energy in 1984, Geothermal Resources Council Bulletin, Vol.14, No.8, pp.3-13, 1985.
[8.3.4] Japan's Sunshine Project, Summary of Geothermal Energy R&D, Ed. by Sunshine Project Promotion Headquarters, AIST-MITI, Pub. by Japan Industrial Technology Association, Tokyo, pp.38-45, 1988.
[8.3.5] Mori, Y., and Suyama, J., Geothermal Energy, OHM Pub. Co., Tokyo, pp.171-173, 1980 (in Japanese).
[8.3.6] Yamada, Y., Types of Geothermal Power Generation, Handbook of Geothermal Energy Development, Chap.8, Sec.2, pp.785-801, Fuji Techno-Systems Pub., Tokyo, 1982 (in Japanese).
[8.3.7] Japan's Sunshine Project, The Symposium Proceedings of R&D Results of Geothermal Energy, Ed. by AIST-MITI, Pub. by Japan Industrial Technology Association, Tokyo, pp.177-216, Sep., 1979 (in Japanese).
[8.3.8] Yamada, Y., and Mori, Y., Pinch Temperatures in Rankine Cycle Power Generation Utilizing Heat Sources of Medium and Small Temperature Differences, Heat Transfer Japanese Research, Vol.14, No.3, pp.55-83, 1985.
[8.3.9] U.S. ERDA Report, SCRC-CR-77-48, Feasibility Demonstration of the Sperry Down-Well Pumping System, July 1977.
[8.3.10] Nonoguchi, M., Summary of 1989 GRC Annual Meeting, Geothermal Energy, Vol.14, No.1, pp.47-53, 1989 (in Japanese).

[8.3.11] Mori, Y., Research and Development of Down-Hole Pumps in the Sunshine Project in Japan, J. Japan Society of Geothermal Energy, Vol.11, No.2, pp.97-108, 1989.

[8.3.12] Kimura, S., Yoneya, M., Ikeshoji, T., and Shiraishi, M., Heat Transfer to Ultra-large-scale Heat Pipes Placed in a Geothermal Reservoir (1st report), J. of Geothermal Research Soc. Japan, Vol.9, No.1, pp.19-30, 1987 (in Japanese).

[8.3.13] Morita, K., Matsubayashi, O., and Kusunoki, K., "Down-hole Coaxial Heat Exchanger Using Insulated Inner Pipe for Maximum Heat Extraction," Geothermal Resources Council, Trans., Vol.9, Part 1, pp.45-50, 1985.

[8.3.14] Austin, A. L., Lundberg, A. W., The LLL Geothermal Energy Program, A Status Report on the Development of the Total-Flow Concept, Lawrence Livermore Lab. Rept., UCRL-50046-77, 1978.

[8.3.15] Ikeda, T., and Fukuda, M., Hot Water Turbine for Waste Heat Recovery, J. Japan Society of Mechanical Engineers, Vol.83, No.745, pp.1528-1534, 1980 (in Japanese).

[8.3.16] Steidel, R. F., Weiss, H., and Flower, J. E., Performance Characteristics of the Lysholm Engine as Tested for Geothermal Applications in Imperial Valley, Lawrence Livermore Lab. Rept., UCRL-80151, 1977.

[8.3.17] Japan's Sunshine Project, Summary of Geothermal Energy R&D, Ed. by Sunshine Project Promotion Headquarters, AIST-MITI, Pub. by Japan Industrial Technology Association, Tokyo, pp.167-176, 1983.

[8.3.18] Yamaoka, K., Biphase (Total Flow) Geothermal Power Production Systems, Jinetsu, Vol.21, No.2, pp.25-36, 1984 (in Japanese).

[8.3.19] Howard, J. H., Present status and future prospects for nonelectrical uses of geothermal resources, NATO CCMS Report No. 40, Oct., 1975

[8.3.20] Sekioka, M., Planning of Hot Water Utilization, Handbook of Geothermal Energy Development, Chap. 8, Fuji Techno-System Pub., p.999, 1982 (in Japanese).

[8.3.21] Sekioka, M., The status of direct uses of geothermal energy in Japan in 1985 including the effect of alternative energy, Geothermal Energy, Vol.12, No.1, pp.107-118, 1988 (in Japanese).

[8.3.22] Report of Basic Design Study of Small- and Medium-sized Binary Cycle Geothermal Power Plant at Shizuku-ishi Area, NEDO and Engineering Advancement Association of Japan, Tokyo, March 1990 (in Japanese).

Table 8.3.1. Capacity of installed geothermal power plants as of 1985

| Country | No. of Plants | Capacity of Plants (MW) |
|---|---|---|
| United States | 67 | 2777.4 |
| Phillipine | 26 | 1059.3 |
| Mexico | 26 | 700.6 |
| Italy | 39 | 459.2 |
| New Zealand | 16 | 300.6 |
| Indonesia | 7 | 289.3 |
| Japan | 14 | 285.1 |
| El Salvador | 4 | 96.1 |
| Iceland | 7 | 71.3 |
| Kenya | 3 | 45.0 |
| Turkey | 2 | 18.3 |
| Nicaragua | 1 | 20.0 |
| Soviet Union | 4 | 15.6 |
| China | 14 | 14.7 |
| Taiwan | 2 | 4.5 |
| France | 1 | 4.2 |
| Portugal | 1 | 3.0 |
| Greece | 1 | 2.0 |
| Total | 235 | 6166.2 |

Table 8.3.4 Capacities of direct use facilities, annual total heat outputs, and load factors in different countries.

| Country | Capacity of Facilities(MWt) | Annual Total Output (GWt) | Load Factor(%) |
|---|---|---|---|
| Iceland | 889 | 5517 | 71 |
| Hungary | 1001 | 2615 | 30 |
| China | 393 | 1945 | 56 |
| New Zealand | 215 | 1484 | 79 |
| Italy | 288 | 1365 | 54 |
| Soviet Union | 251 | 1056 | 30 |
| Rumania | 402 | 987 | 45 |
| France | 300 | 788 | 30 |
| Japan | 135 | 739 | 63 |
| Turkey | 166 | 423 | 29 |
| United States | 399 | 390 | 13 |
| Total | 4379 | 17309 | 45 |

Table 8.3.2 Schedule of DHP program

| Fiscal Year | 1983 | 1984 | 1985 | 1986 | 1987 | 1988 | 1989 | 1990 | 1991 |
|---|---|---|---|---|---|---|---|---|---|
| Studies of Basic Technologies | Literature Survey, Fundamental Experiments, and Check and Review | | | | | | | | |
| No.1 Machine | | | Laboratory and On-site Test | | | | | | |
| No.2 Machine | | | | | | Laboratory and On-site Test | | | |
| No.3 Machine | | | | | | | | Laboratory and On-site Test | |

Table 8.3.3 Goals of DHP Program

| Item | | No.1 Machine | No.2 Machine | No.3 Machine |
|---|---|---|---|---|
| Nominal Well Diameter | (inch) | 9 5/8 | 9 5/8 | 13 3/8 |
| Actual Well Diameter | (mm) | 224 | 224 | 317 |
| Brine Flow Rate | (t/h) | 50 | 100 | 300 |
| Total Pumping Head | (m) | 300 | 340 | 400 |
| Brine Temperature | (°C) | 140 -170 | 170 - 200 | 140 - 200 |
| Installed Depth | (m) | 400 | 400 | 500 |
| Driving Power | (kW) | 100 | 200 | 500 |

Table 8.3.5 Capacities of direct use facilities, annual total heat output, and load factors in Japan.

| Purpose | Capacity of Facilities(MWt) | Annual Total Output (GWt) | Load Factor(%) |
|---|---|---|---|
| Agriculture | 57.03 | 240.8 | 48.2 |
| Aquaculture | 27.49 | 240.8 | 100.0 |
| Heating & Hot Water Supply | 38.22 | 210.3 | 62.8 |
| Snow Melting | 7.30 | 21.1 | 33.0 |
| Tourist Resort | 3.74 | 20.4 | 62.3 |
| Industry | 0.96 | 5.4 | 64.6 |
| Animal Husbandary | 0.16 | 0.5 | 31.3 |
| Total | 134.90 | 739.3 | 62.6 |

Table 8.3.6 Comparison of relative costs of single-flash, double-flash, and single-flash/binary cycle power generation.

| Item \ Type | Single-Flash Steam Turbine | Double-Flash Steam Turbine | Single-Flash & Binary Cycle |
|---|---|---|---|
| Steam/Brine Production Facility | 65 | 65 | 65 |
| Power Generation Facility | 35 | 51 | 87 |
| Construction Cost | 100 | 116 | 152 |
| Capacity of Power | 100 | 119 | 131 |
| Cost / Power | 1.00 | 0.96 | 1.15 |

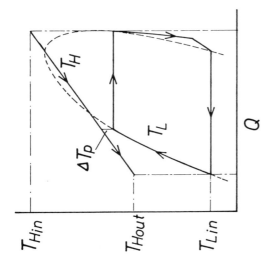

Fig. 8.3.2 Fundamental binary cycle shown as temperature ($T_H$, $T_L$) versus heat exchange rate ($Q$) curves. Brine has the initial temperature $T_{Hin}$, and is wasted at temperature $T_{Hout}$ after heat exchange. The binary medium is heated from its condensing temperature $T_{Lin}$ to evaporation temperature. Saturated medium gas expands in a turbine and condenses in a condenser to make a simple Rankine cycle.

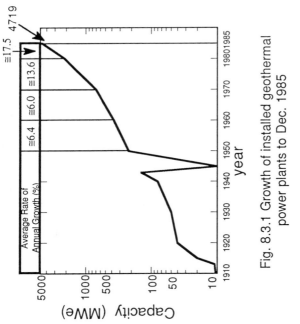

Fig. 8.3.1 Growth of installed geothermal power plants to Dec. 1985

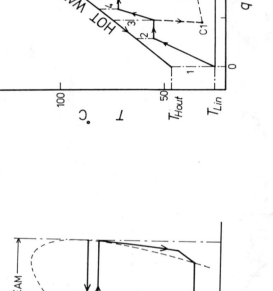

Fig. 8.3.3 Steam-assisted binary cycle. Steam at temperature $T_{Hin}$ evaporates the preheated medium, and brine preheats the condensed medium. The temperature difference between the heating fluid and the medium is kept small throughout the heat exchange process.

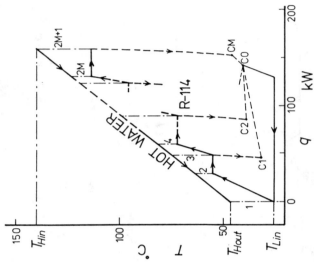

Fig. 8.3.4 Multi-stage evaporation binary cycle showing M evaporation stages. Higher temperature stages have higher evaporation pressures, and the turbine has different stages for medium gas at each stage to expand.

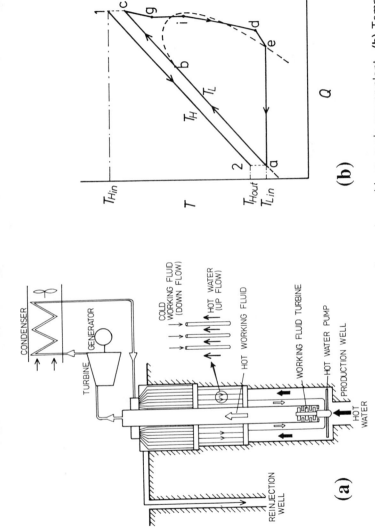

Fig. 8.3.5(a) Schematic of a down-hole heat exchanger/pump binary cycle power plant. (b) Temperature vs. heat exchange curve of the down-hole heat exchanger/pump binary cycle. The binary cycle medium condensed by a condenser at the ground (point a) is sent to a down-hole heat exchanger to be simultaneously heated and pressurized at a super critical state(point c). At the bottom of the heat exchanger, the super critical medium drives a turbine (point c to g) to pump the brine not to flash during rising the well. The medium gas back to the ground drives the power turbine (from point i to d) and condensed (from point d to a) to make a cycle.

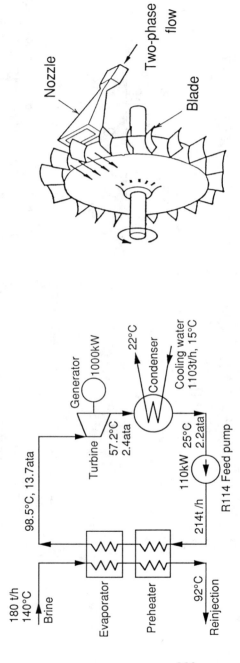

Fig. 8.3.6 Energy flow of a 1 MW binary cycle power plant developed in Japan.

Fig. 8.3.7 Schematic of a total-flow impulse turbine developed by Lawrence Livermore Laboratory.

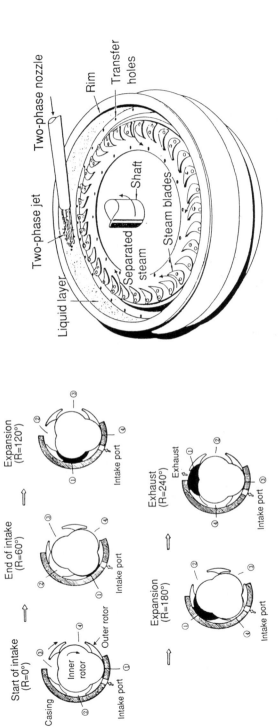

Fig. 8.3.8 Concept of the rotary-type total-flow expander developed in Japan. The inner rotor, the outer rotor and the casing make four spaces (indicated by numbers in circle) for two-phase flow to experience a process consisting of intake, expansion, and exhaust. The inner rotor is placed eccentrically to the outer rotor and the casing. The ratio of the rotations of the inner to outer rotors is 4:3. $R$ in the figure indicates the degree of rotation of the inner rotor.

Fig. 8.3.9 Schematic of the rotary separator of a bi-phase turbine. The two-phase jet from a two-phase nozzle is separated by impinging the inner surface of the rim. The separated liquid gives its momentum to the rotary separator while the separated steam drives the attached blades.

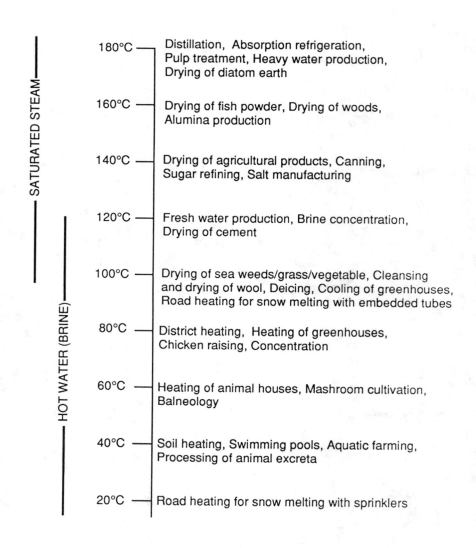

Fig. 8.3.10 Lindal chart showing the various temperature ranges for the direct use of geothermal energy.

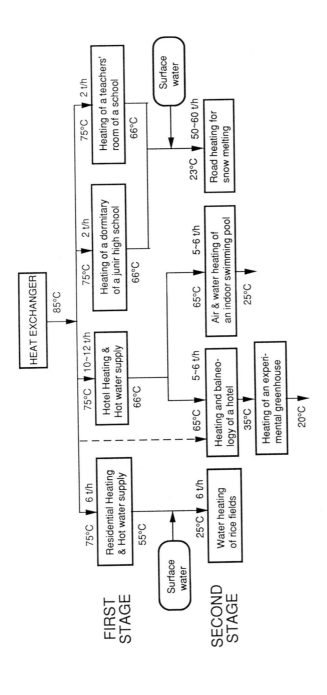

Fig. 8.3.11 Cascading use of geothermal heat at a rural area of Akita in Japan.

# 9. HIGH-TEMPERATURE GAS-COOLED REACTOR AND ITS APPLICATION TO HYDROGEN PRODUCTION

**Yoshiaki Miyamoto**
Department of High Temperature Engineering
Japan Atomic Energy Research Institute
Tokai-mura, Ibaraki 319-11, Japan

9.1 Introduction

In Japan, fossil fuels contributed 84% of the total primary energy supply in 1988, as shown in Fig. 9.1.1. The figure illustrates that 37% of the primary energy supply is converted to electricity as a secondary form of energy. If nuclear energy were employed in the nonelectric applications, 63% of the energy supply, it would be effective in decreasing the release of carbon dioxide produced by burning fossil fuels and would reduce consumption of fossil resources. The ratio of nuclear energy tends to increase to save fossil fuels and to provide a stable energy supply. Presently 38 nuclear power plants produce 24% of the total electrical supply and the ratio of nuclear power generation is projected to increase to 40% by the year 2000.

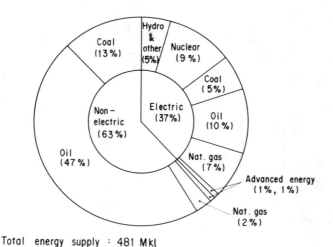

FIGURE 9.1.1. Primary energy supply in Japan ( Based on data from Agency of National Resources and Energy in 1988).

The present-day light water reactor (LWR) and the developing fast breeder reactor (FBR) provide heat at temperatures of 300 and 500° C, respectively, and generate electricity by means of the steam cycle. These are restricted in using heat for nonelectric applications. About 60% of the total energy demands in Japan are for energy having temperatures below 1000° C. The use of nuclear process heat is being considered to meet these demands. The only type of reactor that has the capability to generate electricity and high-temperature process heat is the high-temperature gas-cooled reactor(HTGR). The HTGR can produce 800 to 1000° C heat because its core is composed of ceramic materials and it is cooled by helium gas. Much of the heat energy now provided by fossil fuels could conveniently be supplied by HTGR nuclear heat. Furthermore, if processes to produce hydrogen gas from water with the HTGR can be realized, a complete clean energy system independent of fossil fuels would be established.

Figure 9.1.2 illustrates the level of formation (combustion) energy for carbon, methane and steam. The steam gasification reaction (coal gasification approximately), $C + H_2O \rightarrow CO + H_2$, is endothermic. This heat, a $\Delta H$ of 131 kJ/mol, must be supplied externally. The hydrogasification reaction, $C + 2H_2 \rightarrow CH_4$, is exothermic, $\Delta H$ of 75 kJ/mol. In the figure, the steam decomposition reaction requires the maximum quantity of external heat. The steam reforming reaction, the producer gasification and the steam gasification follow in decreasing order of demand for heat. The producer gasification reaction is not attractive from an environmental standpoint, because the reaction produces two moles of carbon dioxide by combustion of producer gas, two moles of carbon monoxide, into which one mole of carbon is converted. In the case of carbon reactions, the profitable process is to produce the reduced gas by steam gasification and the methane

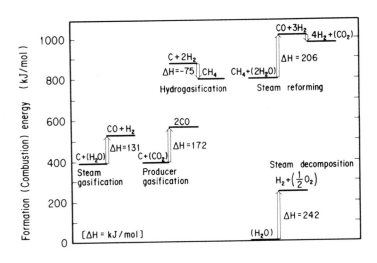

FIGURE 9.1.2. Formation (Combustion) energy for carbon, methane and steam.

gas by hydrogasification. The HTGR is suitable for steam reforming, because this reaction requires the most heat and reacts in the temperature range of 600 to 800° C. The reduced gas obtained by this reaction can be used in various other reactions. These include coal hydrogasification, methanol synthesis, hydrocracking of heavy oil and direct iron making. On the other hand, a number of processes are proposed for steam (and water) decomposition and basic research has been performed on them. Steam decomposition with an HTGR is considered to have potential for thermochemical and electrolytic processes and to be suitable for hydrogen gas production for an industrial plant.

This paper describes features of an HTGR, which has the capability to produce high-temperature heat. It also discusses the nuclear heat process to produce hydrogen gas by typical methods for steam reforming of methane, thermochemical process to decompose water and high-temperature electrolysis of steam.

## 9.2 HTGR Features

An HTGR is a thermal reactor cooled by helium gas and moderated by graphite. The fuel is composed of coated fuel particles, which have good retention of fission products, and a graphite matrix. Uranium oxide and carbide kernels, coated with several layers of pyrolytic carbon (PyC), are uniformly dispersed in fuel compacts. Since graphite material is used as a core body, the HTGR core is different from those of the LWR and the FBR. Features of the HTGR are summarized as follows:

(a) The reactor can operate with coolant temperatures of 800 ~ 1000° C at the reactor outlet, because of the use of inert helium gas and graphite which is stable. At these temperatures, the thermal efficiency exceeds 40% in the steam cycle power generation of HTGRs. Besides electric power generation, HTGRs could be utilized to broaden the applications of nuclear energy to a number of industrial chemical processes and to a gas turbine cycle.

(b) From a nuclear analysis viewpoint, the HTGR core is characterized by a semihomogeneous design, where the fuel particles are dispersed in the fuel compacts. Graphite is used as both the moderator and reflector material because of its small cross-section for thermal neutrons and its excellent structural properties. Studies of various possible fuel cycles, such as a low enriched uranium (U) cycle and a high enriched U-thorium (Th) cycle, have shown the great flexibility of an HTGR.

(c) Even if unusual events like an abnormal power up or reduced core cooling occur, the core temperature will rise slowly because of the large heat capacity of graphite. During an accident in a small modular HTGR core, the decay heat is first conducted to the reactor vessel by natural convection and radiation. Next, it is conducted from the reactor vessel to the environment. In addition, since the temperature coefficient of reactivity is very negative, the reactor shuts down safely.

(d) Helium gas is chemically inert and does not interact with neutrons.

Thus, the primary circuit will have an extremely low level of activity during operation and there will be little radioactive waste.

Fuel

The HTGRs use fuel elements composed of coated particles and graphite and have a prismatic or spherical shape to allow them to be exchanged in the core. Typical prismatic and spherical fuel elements are shown in Fig. 9.2.1.

The HTGR fuel cycle has a number of variants. The reactor can operate on the low enriched U, the U-Th or the Pu-U cycles. The fuel cycle employed in all existing HTGRs utilizes 93% enriched $^{235}$U as the fissile material and $^{232}$Th as the fertile material or low enriched U. In the case of the U-Th cycle, some of the $^{233}$U bred from $^{232}$T is used in situ; that is, it produces fissions together with the $^{235}$U. At the end of a cycle, typically 4 to 6 years, the fuel removed from the reactor still contains a sizable amount of $^{233}$U. This can be either stored or recycled in the reactor.

All HTGR fuels are fabricated using a coated particle concept in which a small kernel, consisting of oxides or carbides of nuclear fuels, is coated with a number of layers of a ceramic material, that is, pyrolytic carbon (PyC) and silicon carbide (SiC). Two types of coated fuel particles have been developed: BISO particles consisting of the fuel kernel, a buffer layer of low-density PyC and an outer layer of dense isotropic PyC, and TRISO particles which are BISO coated with a thin layer of SiC and a dense outer layer of isotropic PyC. The buffer layer can attenuate the recoils of

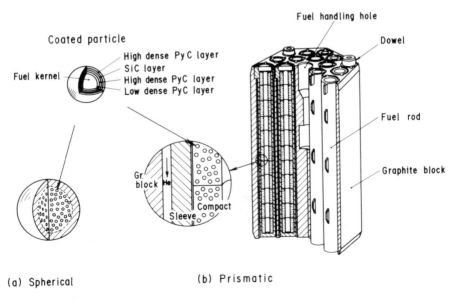

FIGURE 9.2.1. Fuel element of HTGR.

fission products and provide enough void to accommodate fission gases and the swelling of the fuel kernel. An outer layer of dense isotropic PyC provides a high-temperature containment for fission products and mechanical strength for the particles. In TRISO particles, the SiC layer provides a barrier to the release of all metallic fission products, such as Sr and Ba, from the particles at high-temperature operating conditions. An outer layer of dense isotropic PyC provides mechanical support for the SiC layer and acts as a backup containment. Performance of coated particles is generally limited to 1600°C. A burnup of fuel of 100 GWd/t can be permitted, which is higher than that of LWR fuel.

These coated particles are uniformly dispersed in a matrix in the fuel compacts. The matrix consists of an organic binder and graphite filler which is carbonized and heat-treated to yield a carbonaceously bonded compact. The fuel compacts are formed into sticks or cylinders, which are then inserted into graphite sleeves or hexagonal blocks. In the pebble bed reactor, fuel spheres are fabricated by hot pressing after attaching graphite powder to the surface of the particles.

Graphite

In HTGR cores, graphite is used as a structural material, that is, for fuel blocks, reflectors and the core support structure. Graphite has been used in nuclear reactors since their inception, so much information about the properties of graphite for nuclear applications is available. Graphite is stable and has excellent strength at temperatures to about 3000°C. The large heat capacity of graphite makes the transient response of HTGR cores slower than that of LWRs or FBRs during any accident condition. Graphite has little activation, because its small neutron cross-section of absorption. Then fuel and reflector elements discharged can be treated easily. Developments with near isotropic graphites indicate behavior characteristics which are satisfactory for their use in the HTGR, that is, high strength, low anisotropy and minimal change in dimensions due to fast neutron irradiation.

Core and High-Temperature Components

In a core having prismatic fuel elements, the fuel elements are stacked in place by a loading machine. The active core has the approximate shape of a right circular cylinder. It is surrounded by graphite reflector elements which are placed above, below and around the active core. The helium coolant flows downward through the core. Heated helium gas from each fuel channel is mixed in a hot plenum located at the core bottom. Hot gas ducts are connected to the hot plenum to direct hot helium gas to steam generators (SGs) or intermediate heat exchangers (IHXs).

In a pebble bed core, reflector blocks form the plenum, in which fuel spheres (balls) are loaded. The helium coolant flows through the vacuities between the fuel spheres. The spheres are loaded at the top of the core and discharged from the bottom of the core.

The core size of an HTGR is larger than that of an LWR having the same power because of the HTGR's lower power density. A steel reactor vessel is used for HTGRs having a thermal power up to 300 MWt. Similar to the primary circuit on an LWR, this vessel is connected to the hot gas ducts which direct the coolant gas to other components of the primary circuit. An arrangement of a pebble bed core is shown in Fig. 9.2.2. In a large HTGR, the core is contained within the central cavity of a prestressed concrete reactor vessel (PCRV). The PCRV is prestressed by horizontal circumferential tendons and also by vertical tendons. The PCRV ensures an instantaneous failure of the pressure boundary is impossible. An inner carbon steel liner anchored to the concrete provides a gas tight membrane to prevent helium leakage. Water cooling tubes are welded to the concrete side of the liner and a thermal barrier is installed on the reactor side of the liner to protect the concrete against excessive temperatures. Gas ducts in the PCRV are provided to direct helium gas into primary cooling components; the SGs, IHXs and gas circulators. A conceptual arrangement of the prismatic fuel core of a large HTGR is shown in Fig. 9.2.3.

Ceramic insulation is installed on the hot helium gas side of the hot gas ducts and the PCRV to protect the pressure boundary from high temperature. Tube and plate sections, connected to the wall by spacers or attachments, are used to cover the insulation material and to provide a smooth surface for gas flow. Ceramic fiber insulation materials made from $Al_2O_3$ and $SiO_2$, in various forms and design concepts are proposed.

The IHX which transfers the high-temperature heat from the primary to the secondary helium gases, operates at temperature above 900° C. In this application, high-temperature metallic materials such as Incoloy 800H and Hastelloy X are used for heat exchanger tubes and tube sheets. Two arrangements for tube bundles in the IHX have been selected to accommodate the thermal expansion of the heat exchanger tubes: helical coil and U-shape designs. On the other hand, the SG operates at modest temperatures, 400 to 600° C, because the temperature of the tubes and tube sheets of the SG is governed by steam conditions. There are no special considerations for high- temperature conditions in the SG.

Safety

Since the lifetimes of prompt neutrons in an HTGR are 10 times longer than in an LWR, it is difficult for a prompt criticality to occur. Even if a reactivity accident were to take place, the temperature of the core would rise slowly in comparison with an LWR, because of the core with a high heat capacity. To backup the main cooling system, only slow-response auxiliary cooling systems are provided to cool the core. And the core allows long operator response times measured in hours under accident conditions. In contrast, the emergency core cooling system in an LWR needs to start operating within seconds after some primary coolant system failures. Furthermore, a modular HTGR has no provision for auxiliary cooling

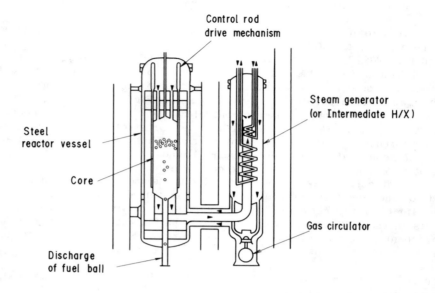

FIGURE 9.2.2. Small and medium HTGR concept with pebble bed core.

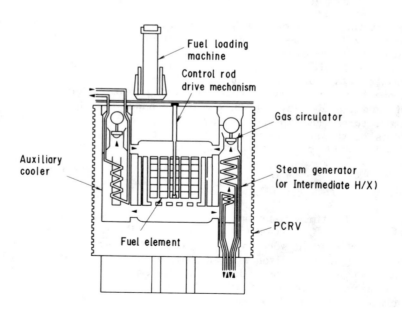

FIGURE 9.2.3. Large HTGR concept with prismatic fuel core.

systems. The activity level in the primary coolant is very low, because helium gas is inert and has very little activation, few fission products are released from the coated particles and the few impurities which are easily removed by the purification system. The radioactive wastes formed during operation are low in an HTGR plant; the volume of solid waste is one-tenth of that of an LWR plant.

## 9.3 Production Processes of Hydrogen Gas with HTGR

Steam Reforming of Methane

Principle. Reduced gas (CO and $H_2$) or hydrogen gas can be produced by steam reforming of methane. The reactions, which involve a nickel-based catalyst, are:

$$CH_4 + H_2O \rightarrow CO + 3H_2 - 206 \text{ kJ/mol} \tag{9.3.1}$$

and under excess steam,

$$CH_4 + 2H_2O \rightarrow CO_2 + 4H_2 - 165 \text{ kJ/mol}. \tag{9.3.2}$$

These reactions are endothermic. The composition of the gas produced by steam reforming depends on the temperature, the pressure and the mole ratio of steam to methane ($H_2O/CH_4$). Gases having the desired compositions are obtained by selecting suitable reaction conditions. With a reaction temperature above 800°C, conversion of methane in the range of 98~99% is achieved. The gas produced is composed principally of hydrogen and carbon monoxide. It is called synthesis gas, because it is used in the synthesis process of methanol, ammonia and so on. In the range of 600 to 800°C, about 70% of methane gas is converted by the process and used for producing lean gas; in the temperature range of 500 to 600°C about 45% of the gas is reformed and utilized for town gas. The conversion is favored by a high ratio of $H_2O/CH_4$, and by high temperature and low pressure. In a steam reformer, it is possible to react carbon deposits which reduce catalyst activity and increases resistance of process gas flow. To prevent carbon formation, steam is added to the methane feed in excess of the stoichiometric quantity.

Conventional system. The conventional steam reforming process has been improved in various ways after it was commercialized in the 1930's. A typical process is shown in Fig. 9.3.1. The feed gas must first be desulfurized to protect the catalysts in the plant. The desulfurized gas is then mixed with process steam and reformed over a nickel-based catalyst in the steam reformer, where the reactions shown in Eqs. (9.3.1) and (9.3.2) occur. The overall reforming reaction is strongly endothermic. It is therefore necessary for the heat needed by the reaction to be supplied to the reaction tubes from a number of burners in the reformer. The process gas from the reformer is cooled to about 350°C and transferred to a high-temperature shift converter, where CO in the process gas is converted to $CO_2$ ($CO + H_2O \rightarrow CO_2 + H_2$). This reaction is exothermic and the temperature increases along the ferrous oxide catalyst bed from 350 to 540°C. The CO content in the process gas is reduced to 2~5 vol% by this

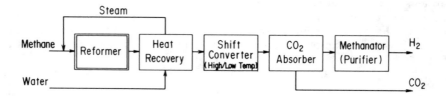

FIGURE 9.3.1. Basic flow scheme for hydrogen production by steam reforming of methane.

reaction. The gas is then cooled to 220° C and it enters the low-temperature shift converter which contains a copper-based catalyst. This shift converter further reduces the CO content in the process gas to under 0.4 vol%. The process gas is then cooled and fed to a $CO_2$ absorber, where more $CO_2$ is removed, attaining a level of 0.1 vol% using chemical absorption methods. The process gas then flows to a methanator, where the remaining CO and $CO_2$ in the gas are methanated as per the reverse reaction of Eqs. (9.3.1) and (9.3.2). The residual CO and $CO_2$ in the hydrogen gas can be reduced to $5 \sim 10$ ppm. The product hydrogen is cooled and sent to downstream units. In a conventional plant, the product hydrogen purity is in the range $97 \sim 98\%$ [9.1].

In the steam reformer, the reaction tubes are heated by burners located in the ceiling, floor and walls. The burner fuel is a light hydrocarbon, LPG or off gas. Since the temperature of the reaction tubes exceeds 900° C, they are fabricated from a high-temperature alloy which has high strength and low creep properties: HK 40 alloy manufactured by centrifugal casting, 25Cr/20Ni steel or Incoloy 800H The typical dimensions of such tubes are: a heated length of $8 \sim 12$ m, an inner diameter of $80 \sim 130$ mm and a wall thickness of $10 \sim 18$ mm. The tubes contain nickel-based catalyst formed rings or pellets. The process gas is reformed, flowing through the tubes.

Hydrogen gas is also produced by the partial oxidation of heavy oil and the gasification of coal, besides reforming light hydrocarbons like methane. In the partial oxidation process, oxygen gas is added to steam and hydrocarbons. The oxygen and hydrocarbons react exothermally and the partial combustion produces the heat required for steam reforming. The typical operating conditions are a temperature of $1300 \sim 1400°$ C and a pressure above 3 MPa.

<u>Steam reforming by nuclear heat</u>. In this adaption of the process, the heat required to induce reforming is supplied by high-temperature high-pressure helium gas from an HTGR core, instead of combustion heat provided for conventional steam reforming. In the conventional system, fuel and preheated air are burned in burners of the steam reformer. The temperature of the combustion gas reaches about 1500° C. The flue gas (combustion gas) flows downwards outside of the reaction tubes. The process gas also flows downwards, but inside the reaction tubes. The reaction tubes contain

catalysts which promote conversion of the gases by the reforming reaction. At the inlet of the tubes, the heat flux is very high because of the large temperature difference between the flue and process gases and the high radiation heat transfer. At the outlet of the reformer, the temperature of the flue gas decreases to about 1000° C and the temperature of the process gas increases to about 800° C. The flue gas discharged from the reformer is used for preheating the process gas and air. Although the film heat transfer coefficient on heating side is small, the average heat flux is comparatively high because of the large temperature difference between the flue and process gases.

In the system adapted for nuclear supplied heat, the maximum temperature of the helium gas is about 900° C. Therefore, it is necessary to induce high heat flux to realize reforming efficiency. The high-pressure helium gas enters the reformer and flows counter to the process gas flow. It is disturbed by turbulence inducing devices to improve the heat transfer performance. Using these techniques, a heat flux can be achieved with a relatively small temperature difference between the helium and process gases. The helium-heated reaction tube has the advantage of a more uniform heat flux around the circumference of the tube. A steam reformer using high-temperature helium from an HTGR has the same performance as a commercial plant [9.2].

These are two systems to couple the reactor and process loops as shown in Fig. 9.3.2: direct coupling which provides hot helium gas directly to the reformer and intermediate loop coupling which provides an indirect one. Direct coupling has the advantages of simple design, low cost and maximum use of the temperature potential. The disadvantages of the direct loop are hydrogen and tritium diffusion in the reformer and some troubles with gas

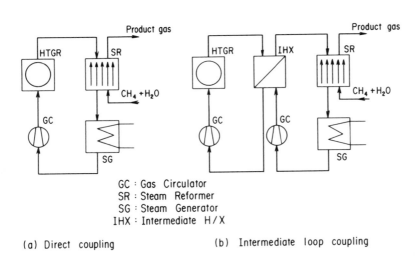

FIGURE 9.3.2. Flow diagrams for nuclear process heat system.

in the event of a tube rupture.

While indirect coupling shares none of the advantages of direct coupling, indirect coupling seems preferable when considering reliability because there have not yet been any instances of a reactor being coupled with a process plant. Indirect coupling with duplex reaction tubes (double tubes) is proposed to provide a buffer zone between the helium and produce gas systems. The buffer zone will be filled with helium gas and monitored to check for leakage between this system and the primary helium system or leakage between this system and the process gas system.

The intermediate loop reduces the temperature of the helium gas actually heating the process gas to about 50° C lower than the primary temperature. If the HTGR core heats the primary helium gas to 950 °C, the helium gas used to heat the process gas will have a temperature of 900° C. The temperature of the helium and other operating conditions change the composition of the produce gas in a steam reformer. Figure 9.3.3 shows the composition of produce gas in a steam reformer for a hydrogen production plant with a capacity of 20,000 Nm$^3$/h [9.3]. The variable parameters are the temperature of the helium gas, the pressure of the process gas, and the ratio of H$_2$O/CH$_4$. Residual CH$_4$ increases with decreasing the ratio of H$_2$O/CH$_4$ and the temperature of the helium gas. With a helium gas temperature of 880° C(case 1) and 780° C (case 2), for example, compositions of 72% H$_2$ and 7%

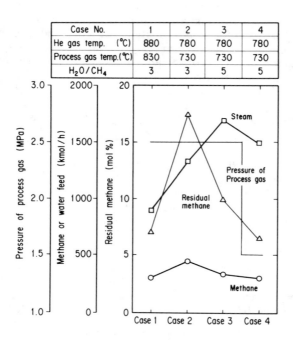

FIGURE 9.3.3. Composition of produce gas in a steam reformer and operating conditions.

residual $CH_4$ for case 1 and 65% $H_2$ and 17% residual $CH_4$ for case 2 are obtained. To obtain the same composition as case 1 with the lower temperature of case 2, it is necessary to reduce the pressure of process gas and to increase the ratio of $H_2O/CH_4$, as shown in case 4.

An energy balance of a hydrogen production plant of 20000 $Nm^3/h$ using a nuclear reforming process and having intermediate loop coupling is shown in Fig. 9.3.4. In the plant, the primary helium gas is heated from 400 to 950° C in an HTGR core and flows to an IHX, where 85 MW of heat is transferred to the secondary helium gas. The flow rate of the secondary gas, 77%, enters the steam reformer, which converts process gas having a ratio of steam to methane, 3.5. Before entering the steam reformer, the process gas is preheated to 500° C by the remaining 23% of the secondary helium gas. The secondary helium gas from the steam reformer and the preheater, now at 700° C, flows to a steam generator and returns to the IHX. In the steam reformer, the process gas is converted to product gas composed of 74% $H_2$, 13% CO, 9% $CO_2$, and 4% $CH_4$. The final product gas, 90% $H_2$ and 10% $CH_4$, is obtained through a refining process using a shift converter and $CO_2$ absorber.

In this plant, the utilization ratio of the nuclear energy is about 34% [ (910° C-700° C)/( 910° C-300° C)] , because hot helium gas having a temperature of 700° C to 910° C is used for the process. Figure 9.3.5 illustrates the energy balance for this hydrogen production plant. Simply defined, the plant thermal efficiency, $\eta_1$, is the ratio of output to input energy as follows:

$$\eta_1 = \frac{Q_{H_2}}{Q_N + Q_F + W_i}, \qquad (9.3.3a)$$

where $Q_N$ is the nuclear energy supplied, $Q_{H_2}$ and $Q_F$ are the heat of

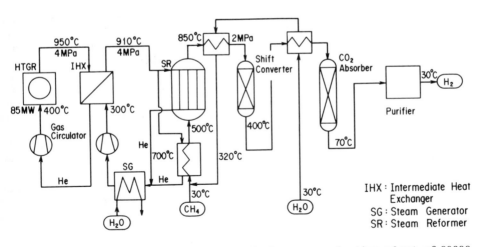

FIGURE 9.3.4. Flow diagram for typical hydrogen production plant of 20000 $Nm^3/h$ using steam reforming of methane.

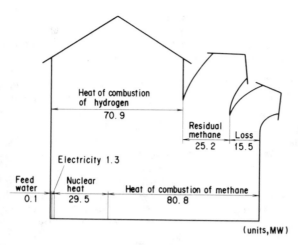

FIGURE 9.3.5. Energy balance of hydrogen production plant of 20000 Nm³/h by steam reforming.

combustion of the hydrogen produced and the feed (methane) gas, respectively, and $W_1$ is the sum of all other input energy. Further, to compare this with other process plants, $\eta_2$ is defined as follows:

$$\eta_2 = \frac{Q_{H2} - Q_F}{Q_N + W_1}. \qquad (9.3.3b)$$

The efficiencies of this plant, $\eta_1$ and $\eta_2$, are 82% and 50%, respectively.

Figure 9.3.6 shows the steam reformer in this plant. The 910° C helium gas enters at the bottom and flows upward in the shell side. During this upward passage, the helium transfers 23 MW of heat energy to the process gas. Then, cooled to 700° C, the helium gas exits at the upper part the steam reformer. Baffle plates are installed in the tube bundle to improve heat transfer performance. Process gas having a temperature of 500° C enters the upper shell and flows downward in the reaction tubes where it is heated to 850° C. The process gas flows from the reaction tubes through connecting tubes to a header. The header is connected to a riser tube, which provides an outlet for the process gas from the steam reformer. There are about 290 reaction tubes. These tubes are fabricated from Inconel 617 and have a diameter of 60 mm, a wall thickness of 5 mm, and a length of 8.4 m. Insulation material is installed on the inner walls of the pressure vessel to minimize heat losses and to prevent an excessive temperature rise of the vessel. These tubes contain catalyst rings having an equivalent diameter of 11 mm. The rings cause the pressure drop of 0.25 MPa in catalyst zone. The helium and process gases have system pressures of 4 MPa and 2 MPa, respectively.

In the steam reforming process utilizing an HTGR, a modified process is proposed to improve the conversion of $CH_4$ to hydrogen gas [9.4]. In this concept, hydrogen gas discharges directly from the reformer tubes through

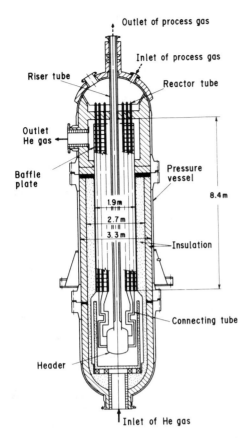

FIGURE 9.3.6. Cross section of steam reformer for hydrogen production of 20000 Nm$^3$/h.

permeable palladium membranes. This increases the efficiency of the chemical reaction and simplifies the process downstream from the reformer.

Processes for Hydrogen Production from Water

Many processes for hydrogen production from water have been proposed. Thermochemical and electrical decomposition are two of them. Though heat at a very high temperature above 4000° C can directly decompose water, it is not possible to generate and control the heat in an engineering process. Thermochemical and electrical decomposition of water both utilize the 1000° C heat generated by an HTGR. These processes are described in this section.

Thermochemical processes. These processes utilize combined chemical reactions which decompose water into hydrogen and oxygen and which convert the reaction by-products for reuse. These processes can be classified into the sulfur-iodine and the iron-halogen cycles by the reactants used in the chemical reactions. A typical well studied process proposed by General Atomic is the sulfur-iodine cycle. It is shown as follows [9.5]:

$$xI_2 + SO_2 + 2H_2O \rightarrow 2HIx + H_2SO_4 \qquad (9.3.4)$$
$$2HIx \rightarrow H_2 + xI_2 \qquad (9.3.5)$$
$$\underline{H_2SO_4 \rightarrow H_2O + SO_2 + 1/2\ O_2} \qquad (9.3.6)$$
$$H_2O \rightarrow H_2 + 1/2\ O_2 \qquad (9.3.7)$$

The main reaction of the cycle is the Bunsen reaction in which iodine reacts with sulfur dioxide to form hydroiodic acid and sulfuric acid in an aqueous solution. The hydroiodic and sulfuric acids then generate hydrogen and oxygen gases, respectively. These latter reactions also generate iodine and sulfur dioxide for reuse in the Bunsen reaction.

When the iodine added to sulfur dioxide is in excess of the equimolar amount, the reaction in Eq. (9.3.4) proceeds to the right side. The solution separates into polyhydroiodic acid (heavier phase) and sulfuric acid (lighter phase) due to the difference in specific gravities of these acids. The separated polyhydroiodic acid solution becomes the mixture $HI/I_2/H_2O=1/4/5$. This mixture can be dehydrated with a dense phosphoric acid which results in separation of the hydroiodic acid from the mixture. The hydroiodic acid can be decomposed thermally in Eq. (9.3.5). The reaction in Eq. (9.3.6) is endothermic. High decomposition of sulfuric acid is obtained at high temperature and under low pressure: for example, the decomposition reaches 69% under conditions of 870°C and 0.4 MPa. The high-temperature heat required for the reaction could be supplied by an HTGR.

Figure 9.3.7 shows a reaction flow scheme of the IS process, an improvement

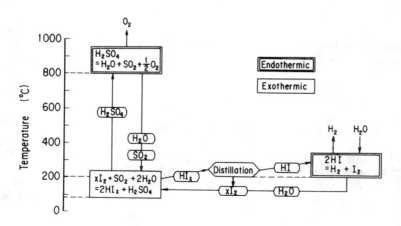

FIGURE 9.3.7. Reaction flow scheme of IS process.

of GA's proposal. The IS process is characterized by separation of polyhydroiodic acid by distillation and decomposition of hydroiodic acid with Pt/activated carbon. The iodine produced is absorbed on the activated carbon which results in a single-pass decomposition having a high efficiency, 60 ~ 70%. The decomposition reaction in Eq. (9.3.5) shifts to the right with increasing absorption of iodine. The absorption depends on kind of activated carbon employed. In an actived carbon column, a saturated absorption zone first forms at the inlet and then proceeds from the inlet to the outlet. After the entire column is saturated by the adsorbed iodine, the activated carbon in the column is regenerated by desorption prompted by hot nitrogen gas. The thermal efficiency of the plant is dependent on the decomposition of the hydroiodic acid and the ratio of steam to hydroiodic acid: high decomposition and a low ratio increase the efficiency.

Figure 9.3.8 shows a flow diagram of a hydrogen production plant having a capacity of 20000 $Nm^3/h$ using the IS process. This plant has a decomposition efficiency of hydroiodic acid, 60%, due to activated carbon and a ratio of steam to hydroiodic acid of 0.43 at the top of the high-pressure polyhydroiodic acid distiller. In the $SO_3$ decomposer, heat from the secondary helium gas at 880° C from the IHX decomposes $SO_3$ into $SO_2$ and $O_2$ in the presence of Pt or $Fe/Al_2O_3$ catalysts. The helium gas then flows to the $H_2SO_4$ decomposer where it dissociates the 95 wt% sulfuric acid into $SO_3$ at a temperature of 400 ~ 500° C. After exiting the $H_2SO_4$ decomposer, the helium gas flows to a steam generator where it generates steam needed by the process. The helium gas returns to the IHX after flowing through another steam generator which provides steam for another purposes. The products, $SO_2$ from the $SO_3$ decomposer and HI from the HI distiller, enter the Bunsen reactor, where a mixed solution of HIx and $H_2SO_4$ is generated and where the acids are separated. The $H_2SO_4$ solution returns to the $H_2SO_4$ decomposer through the $H_2SO_4$ concentrator and the HIx solution separates

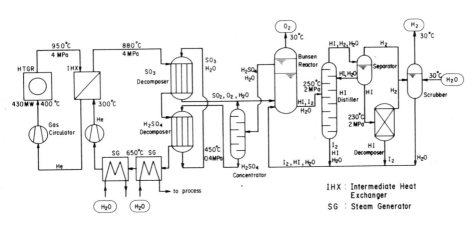

FIGURE 9.3.8. Flow diagram for typical hydrogen production plant of 20000 $Nm^3/h$ using IS process.

into HI, $H_2$ and $I_2$ in the HI distiller. The HI is also decomposed into $H_2$ and $I_2$ in the HI decomposer at a temperature of $200 \sim 300°$ C. Since helium gas within a temperature range of $650°$ C to $880°$ C is used, the nuclear energy utilization ratio is about 40% (($880°$ C-$650°$ C)/($880°$ C-$300°$ C)).

Figure 9.3.9 illustrates the energy balance of the IS process. Nuclear heat from an HTGR provides process heat and generates process steam. The process heat, 107 MW, decomposes sulfuric acid as expresses in Eq. (9.3.6), and the remaining heat, 63 MW, generates process steam, some of which is used to drive process compressors. The compressors have a thermal to mechanical energy conversion factor of 30%. The plant thermal efficiency estimated using Eq.(9.3.3b) is 42% [9.6].

The UT-3 process has been proposed as a practicable representative of the iron-halogen cycle [9.7]. This process employs a Fe reactor and a Ca reactor as shown in Fig. 9.3.10. It utilizes both gas and solid phases, however, in this scheme only the gaseous reactants circulate through the series connected bed reactors. Each reaction is proportional to the partial pressure of the reactant gas, so high pressure enhances the reaction velocity. The plant consists of four pairs of Fe and Ca reactors to keep the cycle operating continuously. The thermal efficiency is expected to reach $40 \sim 45$%.

Hydrogen may also be produced by a hybrid method which combines electrolysis with a thermochemical process. In a typical process, as proposed by the Westinghouse Electric Corp., the following electrolytic reaction and Eq. (9.3.6) are combined [9.8],

$$2H_2O + SO_2 \rightarrow H_2 + H_2SO_4 . \qquad (9.3.8)$$

The theoretical voltage is 0.17 V in this reaction. Although electric power is required, less is needed than required by conventional electrolysis. The

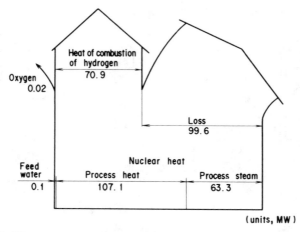

FIGURE 9.3.9. Energy balance of hydrogen production plant of 20000 $Nm^3$/h by IS process.

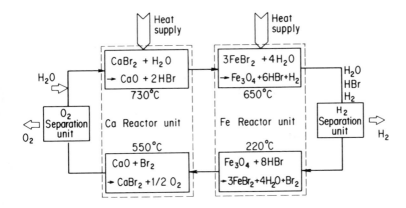

FIGURE 9.3.10. Basic flow scheme for hydrogen production by UT-3 process.

theoretical voltage needed to decompose water is 1.23 V and many commercial electrolyzers require over 2.0 V. The process is potentially capable of achieving a thermal efficiency of 40 to 60%.

High-temperature electrolysis of steam. The high-temperature electrolysis process decomposes steam using 1000° C heat from an HTGR and electricity. The reverse reactions of the solid oxide electrolyte fuel cell are utilized in the process: an oxygen ionic conductor is used as the solid electrolyte, as shown in Fig. 9.3.11. Steam is dissociated with electrons from externally provided electricity on the surface of the cathode. Hydrogen molecules form on this surface. Simultaneously, oxygen ions migrate through the electrolyte and form oxygen molecules on the surface of the anode with the release of electrons. The products, hydrogen and oxygen, are separated by the gastight electrolyte.

The electrolysis cell has an electrolyte of yttria-stabilized zirconia sandwiched between the porous cathode and anode. The cathode is made of a nickel cermet with zirconia or a nickel oxide. The anode is formed from modified lanthanum manganese (or cobalt) oxide. An electrolysis element is composed of several electrolysis cells connected to form a series electrical circuit. For example, the electrolysis element developed by Dornie GmbH (West Germany) forms a cylinder composed of 10~20 cells; each cell having a length of 1 cm and a diameter of 1.4 cm. Flowing inside the element, high-temperature steam is decomposed by electricity. Hydrogen gas is discharged from an exit inside the element and oxygen gas from an exit outside the element. The HOT ELLY (High Operating Temperature Electrolysis) program has demonstrated functional modules composed of many elements connected electrically in series and parallel. A bench-scale electrolysis module which consists of 1000 cells (100 elements with 10 cells) has been assembled. It has a capacity of 0.6 $Nm^3H_2$/h when supplied with about 2 kW of electricity [9.9].

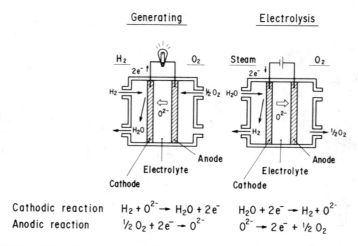

FIGURE 9.3.11. Principle of high-temperature electrolysis of steam.

The total energy demand ($\triangle H$) for steam decomposition is the sum of the Gibbs energy ($\triangle G$) and the heat energy (T$\triangle$S). The electrical energy demand, $\triangle G$, decreases with increasing temperature as shown in Fig. 9.3.12; the ratio of $\triangle G$ to $\triangle H$ is about 93% at 100° C and about 70% at 1000° C. The reaction at the higher temperature provides the advantage in improving the efficiency. Futhermore, improved reaction kinetics at elevated temperatures reduce electrode overpotentials. There are two operational conditions for a plant system using this process: endothermic and exothermic [9.10]. The endothermic operation requires minimum electricity and heat energy to discompose steam supplied to the cells. The exothermic operation requires that only electricity is supplied. Excess electrical energy produces the heat energy required. The efficiency of the exothermic process is lower than for the endothermic operation but the exothermic plant is simpler.

Figure 9.3.13 shows a flow diagram of a hydrogen production system having a capacity of 20000 Nm$^3$/h. An HTGR with a thermal output of 203 MW produces the 64 MW of electricity required to decompose the steam with a power generation efficiency of 38% and 19 t/h of high-temperature steam in the other helium gas loop. The utilization ratio of the nuclear energy is 100% in this process, because the temperature of the utilized energy ranges from 290° C to 900° C. Since exothermic operation is assumed in the process, the plant thermal efficiency is about 35%. The efficiency is expected to be increased to over 40% by the adoption of endothermic operation and the use of electricity generated by high efficiency methods, such as by a gas turbine cycle.

9.4 Conclusion

This paper has outlined that an HTGR has extremely safe characteristics

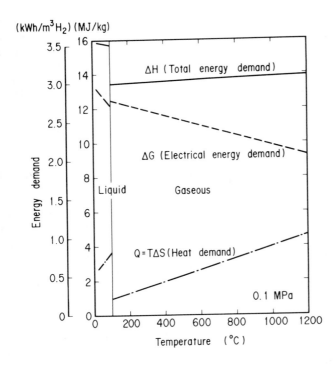

FIGURE 9.3.12. Energy demand for $H_2O$-electrolysis.

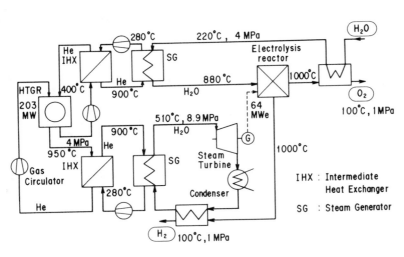

FIGURE 9.3.13. Flow diagram for typical hydrogen production plant of 20000 $Nm^3/h$ using high-temperature electrolysis steam.

and how the HTGR has the potential to meet process heat needs. It also has remarked the system of nuclear heat process coupled to the HTGR. Regarding nuclear heat applications, it has discussed the principle of reaction to produce hydrogen gas and three flow diagrams for hydrogen production plants using different processes: steam reforming, thermochemical process and high-temperature electrolysis. Rough estimation for heat balances of these plants indicates the thermal efficiency in the range of 35～50%.

An HTGR for the steam cycle power generation utilizes heat at 750° C. Two prototype HTGRs have already been constructed and operated in West Germany and the U.S.A.. In Japan, construction of the HTTR (High Temperature Test Reactor) will be started in 1990. The reactor is expected to upgrade HTGR technology and to demonstrate process heat applications for nuclear energy. Regarding a hydrogen production plant with an HTGR, the steam reforming process for methane has the advantage to realize easily coupling with the reactor and will provide plant operation experience in the short term. This is especially attractive because the process is already commercialized and has few technical problems. The only significant issue is the resolution of differences in licensing regulations between a nuclear reactor and a commercial chemical plant. From the long term viewpoint, it is important to continue research on the process for hydrogen production from water.

I wish to thank Mr. S. Shimizu of Department of Research at Takasaki Radiation Chemistry Research Establishment in JAERI for his many helpful suggestions to this paper.

REFERENCES

[ 9.1 ] Balthasar, W., and Hambleton, D. J., Industrial Scale Production of Hydrogen from Natural Gas, Naphtha and Coal, Int. J. Hydrogen Energy, vol. 5, pp. 21-31, 1980.

[ 9.2 ] Kugeler, K., Kugeler, M., Niessen, H. F. and Hammelmann, K. H., Steam Reformers Heated by Helium from High Temperature Reactors, Nucl. Eng. & Design, vol. 34, pp. 129-145, 1975.

[ 9.3 ] CHIYODA Co., Economic Assessment for Steam Reforming with Nuclear Energy, Report to JAERI (in Japanese), 1986.

[ 9.4 ] Oertel, M., et al., Steam Reforming of Natural Gas with Integrated Hydrogen Separation for Hydrogen Production, Chem. Eng. Technol, vol. 10, pp. 248-255, 1987.

[ 9.5 ] Russell, J. L. Jr, et al., Water Splitting-A Progress Report , Proceedings of the 1st W.H.E.C., vol. 1, 1A, 1976.

[ 9.6 ] CHIYODA Co., Study of Flow Sheet for IS Process of Thermochemical Hydrogen Production, Report to JAERI (in Japanese), 1987.

[ 9.7 ] Yoshida, K., et al., A Simulation Study of the UT-3 Thermochemical Hydrogen Production Process, Proceedings of the 7th W.H.E.C., vol. 2, pp. 831-842, 1988.

[ 9.8 ] Brecher, L. E., Spewock, S. and Warde, C. J., The Westinghouse Sulfur Cycle for the Thermochemical Decompositio n of Water, Proceedings of the 1st W.H.E.C., vol. 1, 9A, 1976.

[ 9.9 ] Doenitz, W. and Erdle, E., Recent Advances in the Development of High Temperature Electrolysis Technology in Germany, Proceedings of the 7th W.H.E.C., pp. 65-74, 1988.

[ 9.10 ] Quandt, K. H. and Streicher, R., Concept and Design of a 3.5 MW Pilot Plant for High Temperature Electrolysis of Water Vapor, Int. J. Hydrogen Energy, vol. 11, no. 5, pp. 309-315, 1986.

# 10. ADVANCED ENERGY CONSERVATION

## 10.1 ENERGY-SAVING TECHNOLOGIES IN THE ELECTRIC POWER INDUSTRIES IN JAPAN

**Yukiyoshi Hara**
Chief Researcher, Energy Conversion Department
Engineering Research Center, Engineering R&D Administration
Tokyo Electric Power Company
2-4-1, Nishi-tsutsujigaoka, Chofu, Tokyo 182, Japan

### 10.1.1 Introduction

From 1950's to 1980's, in Japan, the growth rate of electric power demands has surpassed that of Gross National Products (GNP) with some exceptions. Through that period, most of electrical power energies were supplied by thermal power generation and Japan has experienced, so called, oil crisis twice.

From these background, the energy-saving means directly improvement of thermal efficiencies in the power generation. Many efforts have been devoted by both government and industries to improve efficiencies of thermal power stations. These efforts have resulted in the improvement of steam conditions for higher temperature and higher pressure, scaling up of unit capacity, development of sliding pressure type boiler and so on.

However, in recent years, it is considered rather difficult to improve thermal efficiency over the present level because of the material limitation due to high operational temperature. On the other hand, according to the estimation for future electric demands, the demands will continue to increase year by year in Japan and it is predicted that more than half of the total power generation will be generated by thermal power generation. So it is considered that more efforts must be made to improve thermal efficiency of thermal power generation. In addition, from the viewpoints of effective coal utilization, the developments of new power generation systems, such as integrated coal gasification combined cycle system (IGCC), and pressurized fluidized bed combustion system (PFBC), are also considered to be necessary.

In 1980's, Liquefied Natural Gas (LNG) fired Combined Cycle Plants have been introduced in power generation in Japan. The first combined cycle power plant introduced in Japan was No. 3 Group of Higashi-Niigata Power Station of Tohoku Electric Power Company in 1984 (Total capacity 1,090 MW). Next was No. 1 and No. 2 Group of Futtsu Thermal Power Station of Tokyo Electric Power Company (TEPCO) in 1985 (Total capacity 2,000 MW).

These systems have demonstrated surprisingly high thermal efficiencies (approximately 43% on HHV basis) and excellent operational characteristics such as easy start/ stop and load adjustment capability. So, it is expected that Combined Cycle System will take the principal post in power generation in near future in Japan. Since the advance in gas turbine technology are largely responsible for performance gains with combined cycles, many researches and developments for the gas turbine technologies have been conducted to improve thermal efficiencies.

At the same time, the development for Fuel Cell Power Generation System has been, rapidly and widely, conducted in Japan. To this new type of power generation system, many advantages for performance characteristics such as higher thermal efficiency, availability of local power resources, environmental advantages are expected in Japan.

In this section, from the above-mentioned viewpoints, the following subjects will be discussed.
- The history of development for energy-saving technologies in electric power industry in Japan.
- Perspective of the development for recent gas turbine technologies and fuel cell power generation.

### 10.1.2 Improvement of Thermal Efficiency

10.1.2.1 Improvement of Steam Conditions in Conventional Steam Generation.

Figure 10.1.1 shows the improvements of steam conditions, thermal efficiency and unit capacity in case of TEPCO's thermal power stations.

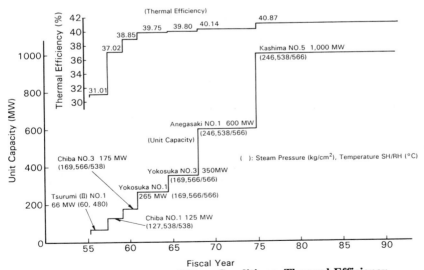

**Figure 10.1.1** Improvements of Stem Conditions, Thermal Efficiency and Unit Capacity in Case of TEPCO's Thermal Power Stations

The following are especially worth mentioning here.

    Tsurumi II-1;
        (1955-1, 66 MW, 60 kg/cm$^2$, 480°C)
        First unit type boiler.
    Chiba-1;
        (1947-4, 125 MW, 127 kg/cm$^2$, 533°C/538°C)
        First reheat type unit
    Yokosuka-1;
        (1960-10, 265 MW, 169 kg/cm$^2$, 566°C/566°C)
        First Cross-Compound shaft unit
    Anegasaki-1;
        (1967-12, 600 MW, 246 kg/cm$^2$, 538°C/566°C)
        First super-critical pressure boiler
    Kashima-5;
        (1974-9, 1,000 MW, 246 kg/cm$^2$, 538°C/566°C)
        First 1,000 MW unit
    Minami-Yokohama-1;
        (1970-5, 350 MW, 169 kg/cm$^2$, 566°C/566°C)
        First LNG only firing boiler
    Hirono-1;
        (1980-4, 600 MW 70 - 246 kg/cm$^2$, 538°C/566°C)
        First sliding pressure boiler

These improvements have been materialized by not only improvement of materials, manufacturing and constructural technology, but also development of supporting technology such as architectural and civil engineering.

10.1.2.2 Necessity of Load Control Ability for the Thermal Power Generation.

The scale-up of unit capacity in thermal power generation has been brought by the necessity to cope with the rapidly increasing electric demand as mentioned in the previous sub-section.

However, a great deal of change has occurred in environment for power generation system in 1970's.

Nuclear power generations were put to practical use and entered into competition as a new power resource.

Also, the difference of power demand between day and night has been spread after the middle of 1970's.

In these circumstances, thermal power generation has been required of the load adjustment capability.

Figure 10.1.2 shows the TEPCO's daily load curve of August 23rd, 1988 on which the maximum demand of this year was recorded.

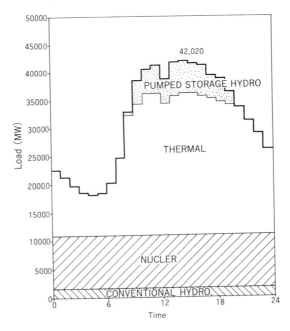

**Figure 10.1.2  Daily Load Curve at TEPCO (Aug-23-1988)**

In this day, the maximum demand was 42,020 MW and the minimum was approximately 18,000 MW. The difference between maximum and minimum demands was approximately 24,000 MW and the fluctuation in demand has been controlled by adjusting the load of thermal power generation.

The sliding pressure type boiler has been developed for this necessity. This type of unit has higher thermal efficiency in partial load, quick response for load change and good operatability such as daily start/stop (DSS) operation.

The reconstruction of existing constant pressure boiler to sliding pressure boiler has been performed for the existing thermal stations and in case of TEPCO the modification has been completed up to some of 600 MW class boilers.

Similarly the expectations for combined cycle plants as future power generations are based on these operational advantages, that is, high thermal efficiency and good load adjustment capability.

10.1.2.3  Trend of Thermal Efficiency

Figure 10.1.3 shows the trend curve of total thermal efficiency (at generating end) in case of TEPCO from 1965 to 1988.

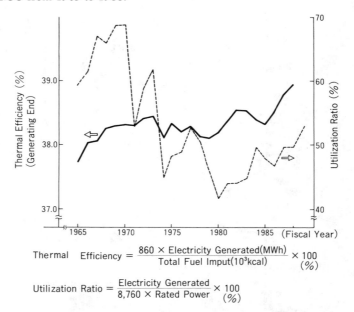

$$\text{Thermal Efficiency} = \frac{860 \times \text{Electricity Generated(MWh)}}{\text{Total Fuel Imput}(10^3 \text{kcal})} \times 100 \; (\%)$$

$$\text{Utilization Ratio} = \frac{\text{Electricity Generated}}{8{,}760 \times \text{Rated Power}} \times 100 \; (\%)$$

**Figure 10.1.3  Trend of Thermal Efficiency and Utilization Ratio in TEPCO**

Although thermal efficiency is directly affected by the unit utilization ratio, the upward trend can be recognized in thermal efficiency from this curve in general.

This is very important and worth mentioning.

These improvements have been achieved not only by the engineering developments but also by the accumulation of minor operational and technical efforts.

Typical examples of these efforts are as follows.

(1) Technical efforts

    ① Replacement of turbine rotor;

        Change from old designed to newly designed rotor as a part of measures for aging degradation

    ② Reconstruction of air heater seal;

        Reduction of air leakage around the air heater by utilizing carbon seal or sensor drive unit.

(2) Operational efforts

    ① Sliding pressure operation;

        In drum type boiler (that is constant pressure boiler), effective means for higher efficiency in partial load conditions.

    ② Minimum $O_2$ of exhaust gas operation;

        Reduction of air flow.

In addition to the efforts mentioned above, the reduction of auxiliary power is also considered important and many efforts have been made in all thermal power stations.

In these periods, there were some circumstantial changes in thermal power generation.

The fuels used in the thermal power generation have been changed from coal, used initially, to distilated oil, crude oil, naphthau and to LNG.

Measures for environmental problem have been requested seriously recently.

The low sulfur contentive fuel oils have been adopted for SOx protection and finally effective de-sulfurization system have been developed.

Several kinds of NOx reduction systems, such as low NOx combustion technology and selected catalytic reduction (SCR) systems have been developed.

As for the dust protection, it must be memorized that first electrostatic precipitator for oil only firing boiler was adopted in Japan.

### 10.1.2.4 Characteristics Required for Thermal Power Generation in Future.

In Japan, electric power demand is anticipated to increase much more in future.

To cope with this increasing demand, TEPCO is planning earnestly to introduce several new power resources for the future from long range viewpoint.

In planning the future power generation, the characteristics required for thermal power generation are defined as follows.

(1) Higher thermal efficiency

    This means good efficiency not only at the rated load but also at the partial load.

(2) Better load adjustment capability

Higher response for the load changing is required in every load, at least 5% per minute over half load to full load zone is required.

Lowering of the minimum operable load is also desired and finally the ability of daily start/stop is also required.

(3) Unit capacity

As far as the unit capacities of thermal power generation plants are concerned, larger is better. Considering that total demand is 49,300 MW in the summer of 1990 in case of TEPCO, unit capacity must be larger.

But in future, small and locally dispersed power resource such as fuel cell power generation will be available from the stand point of stability of total power system and of its higher thermal efficiency.

(4) Environmental aspect

Further improvement of technology is required to reduce the emission of NOx, SOx and dust.

In addition, the protective technology for $CO_2$ emission must be considered in future.

The cost for environmental measures is desired to be less expensive to be used widely.

### 10.1.3 High Efficiency Technology of Gas Turbine

#### 10.1.3.1 Present Status of Combined Cycle Power Plants in Japan

(1) Presently in Japan, a total of 3650 MW is commercially produced through combined cycles at three thermal power stations, including Futtsu No. 1 and No. 2 (total 2000 MW, gas turbine 14 units, steam turbine 14 units) of TEPCO, Higashi Niigata No. 3 (1090 MW, gas turbine 6 units, steam turbine 2 units) of Tohoku Electric Power Company, and Yokkaichi No. 4 (560 MW, gas turbine 5 units, steam turbine 5 units) of Chubu Electric Power Company. And two more combined cycle power plants, Shin-ohita No. 1 and No. 2 (total 1650 MW, gas turbine 12 units, steam turbine 12 units) of Kyushu Electric Power Company and Yanai No. 1 and No. 2 (total 1400 MW, gas turbine 12 units, steam turbine 12 units) of Chugoku Electric Power Company, are under construction or planned to be built.

Table 10.1.1 summarized the combined cycle power plants presently in operation and under construction or planned to be built in Japan.

Turbine inlet temperature (TIT), i.e. temperature of combustion gas at combustor exit, is the key factor to improve the thermal efficiency of the combined cycle plants. The TIT levels of gas turbines utilized in combined cycle plants presently

Table 10.1.1   Japanese Combined Cycle Plants in Service and Planned

### (a) Combined Cycle Power Plants in Service

| Plant | Tohoku Electric Power Co., Higashi-Niigata P.S. No. 3 Group | Tokyo Electric Power Co. Futtsu P.S. No. 1 & 2 Groups | Chubu Electric Power Co. Yokka-ichi P.S. No. 4 Group |
|---|---|---|---|
| Rating | 1090MW | 1000MW x 2 (165MW x 7 x 2) | 560MW (112MW x 5) |
| Plant type | Multi-shaft | Single-shaft | Single-shaft |
| Gas Turbine Type Rating | MW701D 133MW | MS9001E 112.8MW | MS7001E 74.8MW |
| Number of Gas Turbines | 6 | 7 x 2 | 5 |
| Number of Steam Turbines | 2 | 7 x 2 | 5 |
| Fuel | LNG | LNG | LNG |
| Commercial Date | First half; Dec. 1984 Second half; Dec. 1985 | First group; Dec. 1985 to Nov. 1986 Second group; Dec. 1987 to Nov. 1988 | Feb.-July 1988 |

### (b) Combined Cycle Power Plants under Construction or Planned

| Plant | Chugoku Electric Power Co., Yanai P.S. No. 1 & 2 Groups | Kyushu Electric Power Co. Shin-Oita P.S. No. 1 Group | Kyushu Electric Power Co. Shin-Oita P.S. No. 2 Group |
|---|---|---|---|
| Rating | 700MW x 2 (125MW x 6 x 2) | 690MW (115MW x 6) | 870MW (145MW x 6) |
| Plant type | Single-shaft | Single-shaft | Single-shaft |
| Gas Turbine Type Rating | MS7001EA 82.78MW | MS7001E 76.3MW | MW501D 100MW |
| Number of Gas Turbines | 6 x 2 | 6 | 6 |
| Number of Steam Turbines | 6 x 2 | 6 | 6 |
| Fuel | LNG | LNG | LNG |
| Commercial Date | First group Nov. 1990 to Dec. 1992 Second group March 1994 | July 1991 | Three - July 1994 Three - March 1999 |

in service are about 1,150°C and the thermal efficiencies of the plants are nearly 43% (by HHV), which is the highest efficiency level among various power generation methods.

(2) In order to raise the turbine inlet temperature, it is necessary to improve the cooling technology, including manufacturing and fabrication technology, of turbine blades which are made of superalloy.

Recent developments in blade cooling techniques have been achieved along with progress in blade fabrication and heat-resistant materials.

Presently the turbine inlet temperatures of some medium size gas turbines, such as MHI (Mitsubishi Heavy Industries) MF-111 (15 MW) and HITACHI H-25 (26 MW), have already been increased up to 1250 - 1260°C levels.

(3) For the advanced combined cycle plants, the strenuous efforts are being made to raise the TIT to 1350°C-1400°C by improving the cooling technology of turbine blades.

In this case, the introduction of reheat system is also being considered for steam cycle since the temperature of exhaust gas is sufficiently high. The thermal efficiency of advanced combined cycle is expected to be 46 - 47% (by HHV).[10.1.1] [10.1.2]

### 10.1.3.2 Technologies for High Temperature and High Efficiency

(1) Cooling Technologies of Turbine Blades

① In the gas turbines in which turbine blades are made of conventional superalloy system, the increase of TIT is usually accompanied by the increase in the amount of cooling air. However, the conditions necessary to achieve a higher level of turbine efficiency are an increase in the TIT and a reduction of quantity of cooling air flow. Thus, the amount of cooling air consumption must be minimized to achieve maximum benefit of increased turbine inlet temperature. This means that more complex turbine blade cooling technologies are needed to provide acceptable metal surface temperature as the TIT is increased. However, there seem to be a limitation in improving thermal efficiency as long as the air, which is derived from compressor, is used as coolant. Also there is a view in which the present technological level of turbine cooling, utilizing air as coolant, reaches nearly its saturation point.

It seems that the drastic improvement of thermal efficiency can not be expected as long as the conventional air-cooling of superalloy system is utilized.

② Some research efforts are being made for the utilization of steam as coolant because better cooling effect can be expected compared with air cooling due to the larger value of specific heat.[10.1.3] [10.1.4]

Also there are reports in which water could be used as coolant in cooling of turbine blades.[10.1.5]

(2) Application of New Materials

① Introduction of New Processes

A very significant effort is already in progress to replace the conventionally cast alloys with new materials produced by the improved processes such as directionally solidified (DS) superalloys, single crystal (SC) superalloys, and oxide-dispersion-strengthened (ODS) superalloys.

Excellent creep-rupture properties have been achieved through the use of these superalloys. Furthermore, when used with advanced cooling technologies, these superalloys may permit operation at higher metal surface temperature.

② Thermal Barrier Coatings

Thermal barrier coatings (TBC) have been successfully used over the years in applications like combustors and turbine nozzle blades. The application of thermal barrier coatings to rotor blades has been investigated as a part of development efforts of the advanced gas turbine sponsored by the Ministry of International Trade and Industry of Japan. It is reported that the ceramics (yttria stabilized zirconia) coated rotor blades were tested in the high temperature development unit at a turbine inlet temperature of 1400°C and demonstrated about 50°C of heat-resistant effect.[10.1.6]

③ Ceramics

Ceramics, especially silicon carbide and silicon nitride, offers a very attractive combination of high temperate capability (to about 1300 - 1400°C), high stiffness and high strength. This leads to an increase in the usable turbine temperature and the reduction of cooling air. This in turn leads to improved thermal efficiency. Ceramics also offer advantages in terms of wear resistance and erosion/corrosion behavior.

These advantages sparked off a great deal of interest in the early 1970's and a tremendous amount of work has been done in the U.S.A. and Europe since then, especially in the application of ceramics to land-based gas turbines and automobile engines. Two national R & D projects are presently being carried out in Japan to explore small size ceramic gas turbines for automobiles and cogeneration services.

With regard to large land-based gas turbines, there are two groups in Japan presently conducting the research related to the application of ceramics to hot parts of gas turbines. One is a group led by Central Research Institute of Electric Power Industry (CRIEPI)[10.1.7] and the other is the cooperative research program of TEPCO and three Japanese gas turbine manufacturers (Toshiba Corporation, Mitsubishi Heavy Industries, Ltd., and Hitachi, Ltd.). The details of latter project will be describe in the next sub-section.

(3) NOx Abatement Technologies

① The rate of nitrogen oxide (NOx) formation for gas turbines is directly related to firing temperature and dwell time of air in the combustion zone where the temperature is the highest. With increases in turbine inlet temperatures, NOx control and reduction have become more important than ever from the viewpoint of environmental performance and plant efficiency.

② To date the most experience in NOx reduction has been gained in Japan, where extremely low levels of NOx emission standards have been required by governmental regulations. Although NOx emission standard set by government for large land-based gas turbines is 70 ppm at 16% oxygen, the local limits tend to be much lower than this level, especially in case of large power generation plants. For example, the NOx emission limit for the Futtsu thermal power station (2000 MW) of TEPCO is no more than 10 ppm at 15% oxygen, according to the established value agreed upon with the local government.

③ The most effective control technology currently available for reducing NOx emission is to use a combination of steam or water injection in the gas turbine combustion system to reduce thermal NOx formation and selective catalytic reduction (SCR) in the gas turbine exhaust. Using this technology, NOx emissions can be reduced to very low levels in the gas turbine exhaust with no adverse impact on other emissions such as carbon monoxide (CO) or unburned hydrocarbons (UHC). However, the application of this technology has a substantial impact on cost due to added capital investment, operating and maintenance costs associated with the diluent injection and SCR systems. This method has also some impact on the thermal efficiency.

④ Unlike the above-mentioned wet method, dry methods such as two-stage combustion and premixed combustion are effective to improve combustion phenomenon itself and reduce NOx emission without using steam or water injection. The dry methods are presently under development in Japan. Also, the development of an ultra-low NOx combustor is under way which would substantially reduced NOx by catalytically supporting super lean combustion. Much expectation has been placed on the catalytic combustion which is an alternative technology with the objective of obtaining comparable emission levels to the combination of diluent injection and SCR de-NOx system at substantially lower cost.

Conventional combustion process produces a peak temperature which is far into the thermal NOx formation region. In contrast, a catalytic combustion process maintains a peak temperature well below the NOx formation region without compromising combustion efficiency.

Two types of the catalytic combustion concept have been proposed. One employs both homogeneous (gas phase) reaction and heterogeneous (catalyst surface) reaction to accomplish complete combustion within a honeycomb catalyst bed as shown in Figure 10.1.4. It suggests that the catalyst temperature reaches the adiabatic temperature of fuel-air mixtures and that the catalyst should withstand above 1300°C. The other employs only heterogeneous reaction, within a honeycomb catalyst bed, to reduce maximum catalyst temperature as shown Figure 10.1.5. And, complete combustion takes place in a post catalyst bed without a catalyst. The maximum catalyst temperature is maintained around 1000°C.

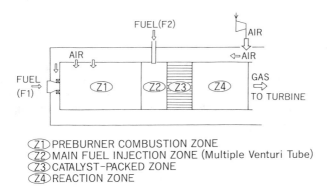

Z1 PREBURNER COMBUSTION ZONE
Z2 MAIN FUEL INJECTION ZONE (Multiple Venturi Tube)
Z3 CATALYST-PACKED ZONE
Z4 REACTION ZONE

**Figure 10.1.4  Catalytic Combustor Model**

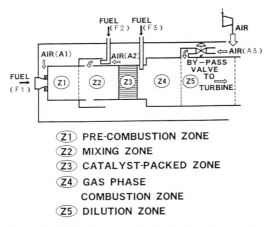

Z1 PRE-COMBUSTION ZONE
Z2 MIXING ZONE
Z3 CATALYST-PACKED ZONE
Z4 GAS PHASE COMBUSTION ZONE
Z5 DILUTION ZONE

**Figure 10.1.5  Hybrid Catalytic Combustor Model**

Based on these concepts, several development programs for gas turbine applications have been conducted in U.S.A. and Japan. They demonstrated very low NOx emission performance (lower than 10 ppmv). Despite the low NOx capability of catalytic combustion, further development efforts are needed in catalyst material, combustor design and control system.

### 10.1.4 Ceramic Gas Turbine

10.1.4.1 Background and Target of the Development

(1) Considering the future energy resources, how to utilize coal energy effectively is an important subject to be solved. Integrated coal gasification combined cycle power generation has been expected as one of the most effective means to use coal energy for future thermal power plant.

(2) Since 1984, Tokyo Electric Power Company has been conducting a cooperative research program with Toshiba Corp., Mitsubishi Heavy Industries, Ltd. and Hitachi Ltd. to develop ceramic hot parts for next generation power generating gas turbine utilized in the coal gasification combined cycle power generation. The improvement of efficiency can be expected by raising turbine inlet gas temperature and reducing the amount of cooling air.

(3) The present goal of the program is the development of 20MW class gas turbine and its turbine inlet temperature is set 1300°C. Ceramics are expected to apply to the combustor, first and second stage stator vane and first stage rotor. As for the first stage rotor, the parallel efforts have been made for the development of thermal barrier coating on the conventional metal blades, because many problems to be solved seem to exist to develop ceramic rotor.

Upon development of these three components, it will be possible to raise the thermal efficiency of existing combined cycle power generation systems from about 43% to about 47%, even if the thermal barrier coated blades are adopted instead of ceramic rotor blades.

The basic specifications of the ceramic gas turbine are shown in Table 10.1.2.

Table 10.1.2  Specification of Ceramic Gas Turbine

| | |
|---|---|
| Output | 20MW |
| Turbine Inlet Temp. | 1300°C |
| Exhaust Gas Temp. | 666°C |
| Pressure Ratio | 15 |
| Shaft Speed | 10800rpm |

### 10.1.4.2 Development Schedule

The development is now at the final stage of the STEP-2 of the component development research, as shown in Table 10.1.3.

**Table 10.1.3  Master Schedule of Development**

|  | 84 | 85 | 86 | 87 | 88 | 89 | 90 | 91 | 92 | 93 | 94 | 95 |
|---|---|---|---|---|---|---|---|---|---|---|---|---|
| Fundamental Research |  |  |  |  |  |  |  |  |  |  |  |  |
| Application Research Development of combustor |  |  | STEP1 |  | STEP2 |  |  |  |  |  |  |  |
| Development of stator |  |  | STEP1 |  | STEP2 |  |  |  |  |  |  |  |
| Development of rotor |  |  | STEP1 |  | STEP2 |  | STEP3 |  |  |  |  |  |
| Material Evaluation |  |  |  |  |  |  |  |  |  |  |  |  |
| Assembly Test of Small Scale Gas Turbine (20MW, 1300°C) |  |  |  |  |  |  |  |  |  |  |  |  |

Given success in the component development research, it is intended to proceed the assembly test, in which a small scale gas turbine using ceramic combustor, ceramic stator vane and thermal barrier coated rotor will be constructed and evaluated. As for the ceramic rotor, the STEP-3 of component research will be continued further.

### 10.1.4.3 Test Results

(1) Combustor

  a. Structure

  As for the liner, 6 stages of ceramic ring were inserted into metal casing. Ceramic insulator was set between ceramic rings and metal casing to keep the temperature of metal below its allowable levels. Ceramic tiles were adopted to reduce the thermal stress for the portion around the air intake holes, where the temperature gradient was steep.

  As for the transition piece, ceramic part were divided into 4 pieces to avoid the generation of excess thermal stresses.

  The structural view of combustor liner and the cross-sectional view of combustor are shown in Figure 10.1.6 and 10.1.7, respectively.

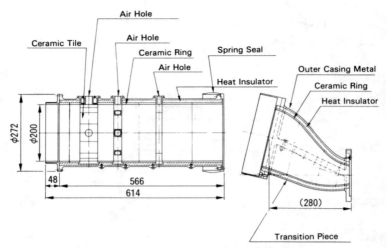

**Figure 10.1.6  Construction of Combustor**

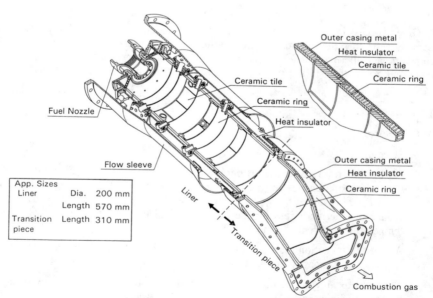

**Figure 10.1.7  Cross-Sectional View of Combustor**

The photograph of combustor linear was shown in Photo 10.1.1.

Photo. 10.1.1  Ceramic Combustor Linear

b. Ceramic material

Pressureless sintered SiC was selected for the ceramic rings and tiles, because SiC has the best properties above 1300°C.

c. Combustion test

Ceramic combustors were tested under the 1300°C and 15ata pressure conditions using pseudo coal gasification fuel.

It was found that high combustion efficiency of more than 99% was obtained over a broad load range, with a pressure loss of 3%, an outlet temperature non-uniformity of about 10% and highly satisfactory NOx exhaust characteristics at the rated load condition.

In these tests, the load-rejection test and trip test were also conducted and it was confirmed that the ceramic parts were healthy and sound after the tests.

(2) Stator vane

a. Structure

A hybrid structure, combining ceramic material and metal, was adopted. Ceramic material of excellent heat resistance was applied to the very hot parts. The ceramics were divided into three, pieces, i.e. an airfoil, inner and outer shrouds considering the reduction of excessive thermal stresses. These were tightened with metal shrouds and a metal core. To reduce the thermal expansion differences between these metal and ceramic parts, Ni-base alloy was applied to the core. Cooling holes were provided at its middle section for internal cooling and thermal barrier coating (TBC) was applied to its surface. Thermal insulation material was inserted between the airfoil and the metal core. Another buffer material was inserted between the ceramic and the metal shrouds. Figure 10.1.8 shows the structural view of the first stage stator vane and Photo 10.1.2 shows the first stage model stator vane.

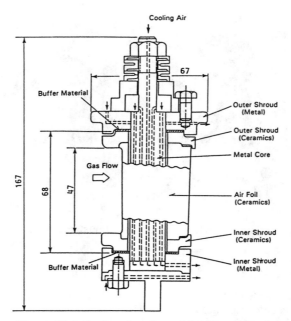

**Figure 10.1.8 Construction of 1st Stage Nozzle**

**Photo. 10.1.2 Ceramic Nozzle (1st stage)**

b. Material

The highest gas temperature at the inlet of stator vane, considering the radial temperature distribution, are about 1400°C in the first stage and about 1100°C in the second stage. Accordingly, pressureless sintered SiC was used for the first stage stator vane, and pressureless sintered $Si_3N_4$ for the second stage. Application of $Si_3N_4$ to the first stage stator vane has also been investigated.

c. Component test

The cascade tests using four sets of startor vanes were carried out under the actual conditions. The cascade used in the tests is presented in Photo 10.1.3. The tests were conducted by the temperature pattern as shown in Figure 10.1.9. The first and second stage stator vanes were sound after a series of tests for steady state and transient (trip) conditions.

Photo. 10.1.3  Nozzle Assembly for Cascade Test

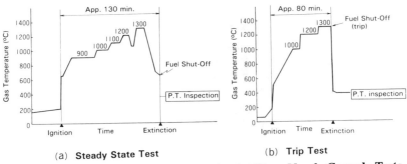

(a) Steady State Test  (b) Trip Test

Figure 10.1.9  Temperature Pattern for 1st Stage Nozzle Cascade Tests

(3) Ceramic rotor

a. Structure

A ceramic blade mounted to a metal shank by a mechanical fitting is planted around the rotor disk of highly reliable superalloy. At STEP-1 of the ceramic rotor development, it was planned to fill a kind of ceramics in this fitting layer.

But, the results of room temperature spin tests showed that ceramic layer could not sustain the sufficient strength. So, at STEP-2, it was changed to metal pads of Ni-base or Co-base alloy inserted symmetrically in this fitting layer to transfer the centrifugal force of blade to the shank, and thermal insulator made of ceramic fiber has been also planned to be inserted into the rest of space of fitting layer to reduce the heat transferred from blades to a rotor disk. Figure 10.1.10 shows the structure of first stage ceramic rotor and Photo 10.1.4 shows its view.

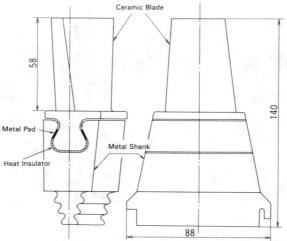

**Figure 10.1.10 Construction of 1st Stage Rotor**

Photo. 10.1.4  Ceramic Rotor (1st stage)

b. Spin test

Tensile tests and room temperature spin tests were performed in order to verify the strength of the ceramic blade subjected to centrifugal force.

At first, tensile tests were carried out at room temperature and 600°C utilizing specimens modeled the fitting part. Then, room temperature spin tests were performed utilizing the simplified 2-dimensional straight blade. Currently, it is confirmed that the present method can sustain the centrifugal force up to the 120% of the rated speed.

In addition to the centrifugal force, rotor is subjected to thermal stresses such as the steady stress at normal operating condition and the transient stress at trip condition. Moreover, vibration problems must be overcomed. To cope with these problems, it is intended to perform the thermal loading test by high temperature combustion gas and the hot spin test at high temperature developing unit, if the specified requirements are attained in the additional test.

(4) Ceramic coated metal blade

a. Structure

The 4-layered thermal barrier coating has been developed in this project. Figure 10.1.11 shows the schematic view of this coating and Photo 10.1.5 shows the coated metal blade. The outermost layer is of zirconia with 8% yttria added and is coated by plasma spraying. The third layer of metal-ceramic mixture reduces the thermal stress within the coating, much more effective than conventional 2-layered coating. The second metal layer is effective to prevent the oxidation of the mixture layer.

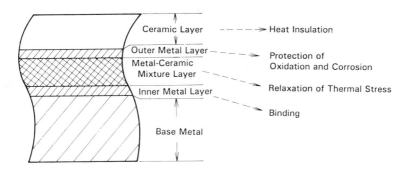

**Figure 10.1.11 Thermal Barrier Coating (Four-Layer Type)**

Photo. 10.1.5 Thermal Barrier Coated Rotor (1st Stage)

b. Thermal cycle test

The results of the thermal cycle tests by burner heating and air cooling show that the life of 4-layered coating is about two times as much as that of 2-layered one.

c. Cascade test

The cascade tests at the actual thermal condition have performed utilizing 4-layered coated blades. The temperature of base metal was measured with and without the coating, and it was found that the coating have about 90°C of heat-resistant effect. This result is superior to ones of 2-layered coating, and is very close to the estimated effect in advance of the test.

In addition, the cycle tests of one hour holding to high temperature followed by the trip was carried out ten cycles to evaluate the endurance of coating, and no apparent damage was detected by the inspection after the test.

Although the long-term test and the test under centrifugal stress field have not yet been performed, there is a strong possibility that 4-layered coating can be practically used to the rotor blades of gas turbine. As mentioned before, it is planned firstly to verify the validity of the ceramic coated metal blade in the research of forthcoming assembly test. After that, it is considered to replace the coated blade by the ceramic rotor in this assembly test, if satisfactory results are obtained in the additional research for ceramic rotor.

## 10.1.4.4 Ceramic Material

In the beginning of this program, $Si_3N_4$ and SiC were regarded as the most promising material for ceramic gas turbine parts. Although the $Si_3N_4$ ceramics showed excellent strength, toughness and resistance to thermal shock below 1200°C, it degraded rapidly above 1300°C. Then the SiC ceramics were considered as the most promising material for combustor, first stage stator vane and first stage rotor, at the expense of mechanical properties.

The operating conditions of the ceramic components are summarized in Table 10.1.4.

**Table 10.1.4  Operating Conditions Ceramic Components (Approximate Value)**

|  | Combustor | Nozzle 1st | Nozzle 2nd | Rotor |
|---|---|---|---|---|
| Gas Temperature (°C) max | 1500 | 1400 | 1100 | 1220 |
| mean | --- | 1300 | 1000 | 1150 |
| Gas Pressure (ata) | 15 | 15 | 6 | 10 |
| Gas Velocity (m/sec) | >50 | >700 | >600 | >600 |
| Promising Material | SiC | SiC | $Si_3N_4$ | SiC |
| Surface Temperature (°C) | 1400 | 1380 | 1080 | 1200 |
| Thermal Stress (MPa) | 60 | 290 | 200 | 400 |
| Centrifugal Stress (MPa) | --- | | | 290 |

Recently, high performance $Si_3N_4$ ceramics have been developed. They demonstrate little degradation of strength and high resistance to oxidation up to around 1400°C.

Although the corrosion resistance in the combustion gas atmosphere and long term reliability at high temperature are not clarified, it is expected to increase the reliability of ceramic components if $Si_3N_4$ can be adopted for first stage stator vane and first stage rotor instead of SiC.

## 10.1.5 Fuel Cell Generation Technology [10.1.8]

Fuel cells are electrochemical devices that convert the chemical energy of a reaction directly into electrical energy. In a typical fuel cell, gaseous fuels are fed continuously to the anode (negative electrode) compartment and an oxidant (i.e., oxygen form air) is fed continuously to the cathode (positive electrode) compartment, the electrochemical reactions take place at the electrodes to produce an electric current.

A fuel-cell stack usually consists of a number of individual cells connected in electrical series, as depicted by the schematic representation of the basic cell structure in

Figure 10.1.12. Each cell contains an electrolyte matrix in contact with a porous anode and a porous cathode. The individual cells in Figure 10.1.12 are connected by a ribbed bipolar separator plate.

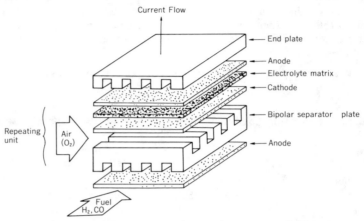

**Figure 10.1.12 Expanded view of basic fuel-cell structure with the repeating unit in a fuel-cell stack**

Three type of fuel cells have been developed for an electric utility, and they are usually classified according to the type of electrolyte used in the cells; (i) phosphoric acid fuel cell (PAFC), (ii) molten carbonate fuel cell (MCFC) and (iii) solid oxide fuel cell (SOFC). These fuel cells are listed in the approximate order of increasing operating temperature, ranging from -200°C for PAFC, -650°C for MCFC and -1000°C for SOFC. In low-temperature fuel cells (PAFC), hydroxyl ions are the major charge carrier in the electrolyte, whereas in the high-temperature fuel cells, MCFC and SOFC, carbonate ions and oxygen ions are the charge carriers, respectively.

From the viewpoint of utility PAFC, MCFC, SOFC are discussed here.

10.1.5.1 Advantages and Possibilities Anticipated as an Electric Utility

(1) High Efficiency:

- a. The current power generating system passes the stage of thermal energy during the conversion process into electrical energy, so it's efficiency obeys Carnal's cycle. Moreover it passes the other stages (chemical energy → thermal energy → mechanical energy → electrical energy), so there is substantial energy loss. In the case of fuel cells, high conversion efficiency can be expected since chemical energy is converted directly into electrical energy.
- b. With the phosphate type, exhaust heat of around 200°C is obtained, 600 - 700°C is obtained with the molten carbonate types, and 800 - 1000°C with the solid electrolyte type. Thus, combination is possible with local heating as cogeneration. Moreover the efficiency of MCFC and SOFC can be increased by combination with bottoming cycles, since steam supply is easily obtainable in these fuel cells.

c. High efficiency can be obtained even by small power stations because efficiency is independent of output capacity.

  d. High efficiency can be expected even at partial load.

  To summarize all of the above-mentioned features, generating efficiency as a complete system is likely to be 40 - 50% for the phosphate type, 50 - 60% for the molten carbonates type, and 50 - 65% for the solid electrolyte type. It can be said that this is ideal as a generating system using fossil fuel.

(2) Superior Environmental Capability:

  a. Air pollutants is expected to be very little.

  b. Noise is reduced due to few mechanical parts in the system, so it is suitable for distribution of the stations in both metropolitan and suburban locations.

(3) Modular-structured Facilities

  Due to the modular structure of cell units, fabrication, transportation and extension are possible, thus providing shortened construction periods as well as installation of the required capacity at the required time. Moreover maintenance and inspection are available as modular units, and replacement for maintenance can be easily performed.

## 10.1.5.2 The Current State of Development and Future Problems

(1) Phosphate Type

  a. Current State of Development:

  Since this is the most developed type, some are of the opinion that operation is now ready for practical use. In the demonstration of 4.5 MW plant test by TEPCO completed in 1985, a power plant achieved 2423 hours of cumulative time and 5,428,240 KWh of accumulated generated output, providing valuable data on efficiency reliability and safety. An 11MW power plant is currently under construction by TEPCO. The running test is scheduled to start 1991 with emphasis on plant performance and reliability.

  b. Problems:

  * Reliability: Some laboratory-scale bench units achieved a life of over 40,000 hours, but 15,000 hours was the longest life for the field test.

  * Cost: The current construction cost of phosphate fuel cell power plant is approximately 1 million yen per KW. Compared with about 2 hundred thousand yen/KW for existing thermal power plant, this is rather expensive. So there is a necessity for drastic cost reductions. For this purpose, it is necessary to simplify the plants, do away with unnecessary equipment by improving reliability and make efforts to reduce costs by mechanization and automation of manufacturing.

(2) Molten Carbonate type

   a. Current State of Development:

   As a national project, a 1KW-class stack was developed in 1984 and a 10KW-class one in 1988. Based on the results, a 100KW-class plant is now under development. The interim assessment will be made in 1991. Based on these results, the 1,000KW-class power plant and system is scheduled to be developed by 1995.

   b. Problems:

   * Durability: Although 6,000-hour durability is required in continuous operation and 40,000-hour durability in cumulative operation, these have not yet been achieved with the exception of small unit cells. The major problem is caused by the deterioration of materials, such as the problem of the cathode dissolving in the electrolyte, the anode and cathode being deformed over time, and the durability of the electrolyte support in the electrolyte.

   * Cost: The cost is expensive at the current level.

   * Total Systemization: A total system combined with bottoming cycles has not yet been operated.

   * Development of Materials: It is necessary to improve the durability of component materials.

(3) Solid Electrolyte type

   a. Current State of Development:

   Compared with the above two systems, this development has a shorter history. The testing machine (MITSUBISHI/TEPCO/ELECTRIC POWER DEVELOPMENT CO., LTD.) achieved the maximum 1.3KW during the period 1989-1990. Continuous-operation performance is now under test. In addition, the 25KW test machine made by WESTINGHOUSE ELECTRIC CORPORATION is scheduled to begin test operations late in 1990.

   b. Problems:

   * Cost: The cost is high, so that simplification and quality control of the process of manufacture is now conducted to reduce the cost. But considering that one of the causes of the high cost is the use of expensive rare earth elements (La, Y, etc.) as electrode materials, it is also desirable to develop electrodes which require a small amount of these elements.

   * Durability: There are several causes for reducing durability such as the reaction between electrodes and electrolytes, the decreases in the active spots and the diffusion paths of electrodes, and the peeling of electrodes and cracking of electrolytes. Various appropriate countermeasures need to be examined for each case.

   * Reliability: Since a great part of the cell materials consist mainly of ceramics, we face the problem of fragility of ceramics. For higher plant reliability, it is desirable to strengthen the ceramics itself.

# Reference

[10.1.1] D.E. Brandt, ASME Paper 87-GT-14 (1987)

[10.1.2] A.J. Scatzo, et al., ASME Paper 88-GT-162 (1988)

[10.1.3] Yanai, et al., Proc. of the 14th Regular Meeting for the Gas Turbine Society of Japan, June 1986.

[10.1.4] "Ten Year's Progress of Moonlight Project: Research and Development of the Advanced Gas Turbine," published by the Engineering Research Association for Advanced Gas Turbines (Japan), 1988.

[10.1.5] Ohno, et al., Proc. of the 14th Regular Meeting for the Gas Turbine Society of Japan, June 1986.

[10.1.6] Imai, OHM, July 1988.

[10.1.7] Abe, et al., Proc. of the 17th Regular Meeting for the Gas Turbine Society of Japan, June 1989.

[10.1.8] Fuel Cells A Handbook (1988) by K. Kinoshita. F.R. Mclarnon & E.J. Cains. U.S. Department of Energy Office of Fossil Energy.

## 10.2 DEVELOPMENT OF "SUPER HEAT PUMP" AND ITS CONTRIBUTION TO GLOBAL ENVIRONMENT PROTECTION

**Akira Yabe**
Mechanical Engineering Laboratory
Agency of Industrial Science and Technology
Ministry of International Trade and Industry
Tsukuba Science City, Ibaraki 305, Japan

### 10.2.1. Outline of "Research and Development of Super Heat Pump Energy Accumulation System" and the effectiveness and environmental characteristics of the heat pump

The "Super Heat Pump" used in the above title refers to the high performance heat pump being developed by the Japanese National energy conservation projecter entitled "Research and Development of Super Heat Pump Energy Accumulation System". This is one of the large scale energy conservation technology research and development projects backed by the Agency of Industrial Science and Technology, Ministry of International Trade and Industry. The leading features of the high performance heat pump are mainly in the following two categories. First are the high efficiency heat pumps which are expected to offer a coefficient of performance (COP) nearly double the conventional level. The other are two high temperature thermal outputs type compressor-driven heat pumps which are expected to achieve high temperature thermal outputs of about 150°C and 300°C. These two types of heat pumps have been called "Super Heat Pump", which means ultra high performance compression heat pumps. In this paper, the results of the interim evaluations for super high performance compression heat pumps have been explained, emphasizing the detailed cycle structure, the development of the higher performance components such as the heat exchangers and the compressors and the contribution to energy conservation and global environmental protection.

Concerning the characteristics and the effectiveness of heat pumps, the important features can be drawn by a comparison of the various methods of heating rooms through the energy flow and the available energy flows. As shown in Fig.10.2.1, the efficiency of the energy conversion would be nearly equal among an electric heater, a gas FF(Forced Flue) heater and a heat pump for obtaining hot air of 50°C. However, as seen from the available energy flow, the efficiency of the available energy conversion would be below 10% for the electric heater and the gas FF heater, and about 40% for the heat pump. Therefore, it could be concluded that the usage of high performance heat pumps would be most effective from the viewpoint of energy conservation and that the usage of the electric heater and the gas FF heater would be qualitatively inadequate, and in a sense wasteful, by considering their ability of realizing the much higher temperature.

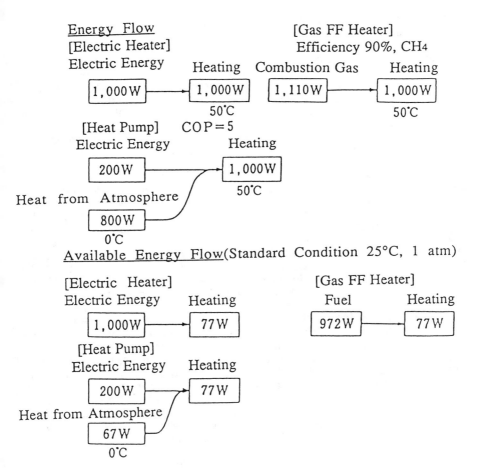

Fig. 10.2.1. Comparison among various air conditioning methods by energy flow and available energy flow (for heating)

The concept of this project is illustrated in Fig.10.2.2. The purpose of this project is to develop the above-mentioned "Super Heat Pump" and chemical heat storage systems for the final development of total systems in which heat is stored in the storage equipment during night time. The stored heat is then taken out in the form of the high temperature heat or the low temperature heat during the day time to be effective for the electric power load levelling, when total energy demand would become maximum because of the usage of air conditioning of buildings, hot water supply, industrial heating, etc. This system, when successfully developed, is expected to reduce electric power required by the high efficiency heat pump for air conditioning purpose, equalize the electric loads through the use of thermal storage, and finally, accomplish remarkable energy conservation.

Concerning the maximum COP of an ideal heat pump, the ideal heat pump cycle could be shown as the dotted line in Fig.10.2.3. For the heating condition of JIS (Japanese Industrial Standard) where there is a temperature increase of 35K from 10°C to 45°C, in the heat pump the heat source fluid enters at 10°C and leaves it at 5°C, and where the heat load fluid enters the heat pump at 40°C and leaves it at 45°C. Then, the maximum COP value estimated on the basis of ideal conditions of heat exchange at zero temperature difference will be 9, to be obtained from COP = 1 / [1- { 278K (heat source fluid outlet temperature ) + 283K (heat source fluid inlet temperature) } / { 313K (heat load fluid inlet temperature) } + 318K (heat load fluid outlet temperature) }. In the super heat pump project, the target COP of the development for the above temperature conditions is 6 and COP of 5.8 has been already achieved for the 100kW class thermal output smaller scale bench plant. Therefore, from the viewpoint of the effectiveness of available energy, the ideal heat pump of COP=9 is 100% effectiveness for the condition where the standard of the available energy would be taken as the heat source temperature, and the super heat pump of COP=6 would become about 70%. This high valve of COP means a lower efficiency of electric power generation of below 10% by utilizing the waste hot water of 45°C. This characteristic becomes clearer for the smaller temperature increase. Therefore, concerning the heat demands of air conditioning and water supply for buildings and residential use, the utilization of heat pumps to raise the temperature and to supply heat would be effective from the viewpoint of energy conservation. As for the utilization methods of waste heat from factories, there would be two main methods, one to generate electricity by use of an electric power plant and the other to raise the temperature of the waste heat so as to supply hot water and industrial heat demands. The selection of the utilization methods has to be based on the quality and quantity of the heat demand. In particular, the temperature of heat demand and the types of demands such as power, electricity and heat, have to be examined totally for the individual energy demand so as to realize the optimum energy system for each factory. This would be composed of the electric power generation, the heat pump and the cogeneration, etc.

Concerning the characteristics of high performance heat pumps for the aspects of environmental protection, two important features will be drawn. The first one is the reduction of local emission of exhaust $CO_2$ and $NO_x$ by use of electricity (the driven compression type) as compared with the boiler system. As is detailed in Chapter 4 of this book, the larger emission facilities of carbon dioxide and nitrogen oxide, such as electric power stations, will probably use

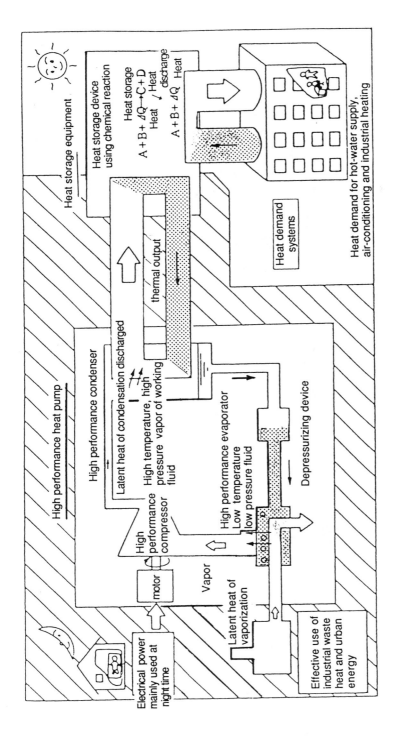

Fig.10.2.2. Schematic diagram of "Super Heat Pump Energy Accumulation System"

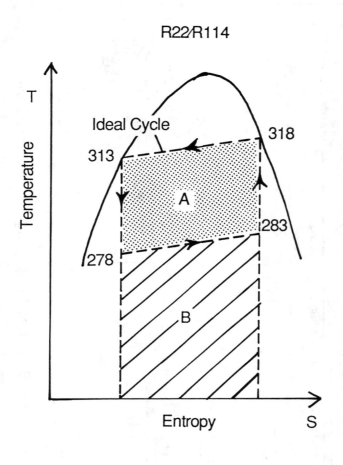

Fig.10.2.3. Ideal heat pump cycle for temperature conditions

the future disposal technology of $CO_2$ and the cleaning technology of exhaust gas. But the small scale combustion facilities have been widely distributed and would be difficult to be equipped with the future disposal technology of $CO_2$ from an economical point of view. Also by considering the energy resources for the electric power generation which could contain water power, nuclear energy and LNG (Liquefied Natural Gas) and which could reduce the emission of carbon dioxide as relatively compared with the usage of oil for combustion facilities, the electrically driven heat pumps could play an important role as one of the environmental protection technologies. The second feature is the reduction of fossil fuel consumption. When the COP of a high performance heat pump is assumed to be 6 and the thermal efficiency of the electric power generation is 38%, 2.3 times the heating value of the fossil fuel could be utilized. It would be clear that the fossil fuel consumption could be reduced below half, even if compared with the boilers 100% thermal efficiency and that the high performance heat pumps would be important environmental protection equipment.

As described above, the necessity and the importance of requesting the higher performance heat pumps are very clear, although the principle of heat pumps had already been established in the mid 19th century. Among many kinds of heat pumps such as an electrically driven compression heat pump, an engine driven compression heat pump, an absorption heat pump, a chemical heat pump, etc, the present conditions of the research and development of the compression heat pump are mainly explained in the following subsections. The chemical heat pump will be described in section 10.4.

### 10.2.2. Interim evaluation of "Super Heat Pump Energy Accumulation System"

A preliminary review of the master plan was conducted in 1989 after five years of research and development for realizing the innovative components, the bench plant of 100kW heat pumps and 10,000kcal class chemical heat storage. The results are shown in Table 10.2.1 and Table 10.2.2. As seen from the tables, nearly all of the bench plants have successfully developed and exceeded the criteria for bench plants. The goals for the pilot plants of super high performance compression heat pumps are shown in Table 10.2.3. As seen from the table, the goals of the actual proof test for wider application have been added to the master plan through discussion of interim evaluation.

In the interim evaluation process, the qualification methodology for high performance compression heat pump systems has been researched. Particular attention has been paid for data fluctuations caused by variations in operational environment, characteristic time of the system, thermal disturbances, and electromagnetic noise from a power supply or control/auxiliary systems. The operation data were collected every 10 seconds for a few hours, and the averaging procedures were examined to eliminate the influences of possible external disturbance with a wide spectrum of frequency, as shown in Fig.10.2.4. Through detailed analysis of the results, the optimum procedure for evaluating the stationary performance of the heat pump system has been developed. The proposed procedure comprises the following steps: (1) averaging 12 series data collected in 2 minutes to form a data set; (2) further averaging 6 such data sets which were collected every 10 to 20 minutes; and (3) calculating performance indicators; such as COP or compressor efficiency, by using the reduced average data.[10.2.4]

Table.10.2.1.
Results of Interim Evaluations for Super High Performance Compression Heat Pumps

| Item | | Type of Compressor | Interim evaluation test for COPs | | Thermal output / kW (Results) | Refrigerant |
|---|---|---|---|---|---|---|
| | | | Criteria | Results | | |
| High efficiency type | For heating | Turbo (4-stage) | 8.0 [8] | 7.7* | 1600 (1520) | R113/R114 (90:10 mol%) →R123/R134a(pilot) |
| | For heating and cooling | Screw (2-stage economizer) | 5.5 [6] (heating) 6.5 [7] (cooling) | 5.8 6.7 | 190 (193) 170 (173) | R22/R114 (87.5:12.5 mol%) →R22/R142b(pilot) |
| High temperature type | For low temp. heat source | Screw (2-stage) | 2.8 [3] | 2.8 | 180 (128) | TFE/H$_2$O (85:15 mol%) |
| | For high temp. heat source | Reciprocating (with a turbocharger) | 2.2 [3] | 2.4 | 300 (208) | H$_2$O |

* The COP=8 will expected to be achieved at the pilot plant.   [ ] :Goals of COP for this project

Table.10.2.2. Results of Interim Evaluations for Chemical Heat Storage Units

| Item | | Output temp. °C | Interim evaluation test | | | | | Total heat storage / kcal (results) |
|---|---|---|---|---|---|---|---|---|
| | | | q: Heat storage capacity | | η: Heat recovery rate | | | |
| | | | criteria | results | criteria | | results | |
| High temp. heat storage | Ammonia complex | 200 | 26 kcal/kg | 37.6 kcal/kg | 20 % | | 38.2 % | 6,000 (8,642) |
| | Hydration | 150 | 41 | 50.0 | 55 | | 72.6 | 11,650 (14,109) |
| | Solvation | 85 | 34.5 | 37.7 | 57.8 | | 60.5 | 11,175 (12,233) |
| Low temp. heat storage | Solute mixture | 7 | 34 | 35.6 | 75 | | 60.8 (82.0) | 31,800 (33,363) |
| | Clathrate | 9.5 | 30 | 31.2 | 62 | | 64.4 | 30,000 (31,264) |

q: Heat storage capacity (kcal/kg)   η: Heat recovery rate (%)   1 cal = 4.19 Joule

Table 10.2.3.
Goals for pilot plant of super high performance compression heat pump (~1993.3).

| Item | | COP (Temperature Charge) | Thermal Output |
|---|---|---|---|
| High efficiency type | For heating | COP=8 (50°C → 85°C)<br>COP=8 (35°C → 65°C)* | about 2,400kW |
| | For heating and cooling | COP=6 (10°C → 45°C)<br>COP=7 (32°C → 7°C) | 1,000kW |
| High temperature type | For Low temp. heat souce | COP≥3 (50°C → 150°C)<br>COP=5 (95°C → 150°C)* | 400kW |
| | For high temp. heat source | COP≥3 (150°C → 300°C)<br>COP=6 (200°C → 300°C)* | 1,000kW |

*Actual Proof Test for Wider Application

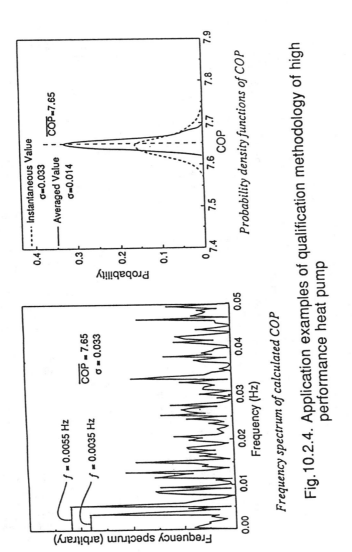

Fig.10.2.4. Application examples of qualification methodology of high performance heat pump

Concerning the development of chemical heat storage systems, the ammonia complex has utilized $NaSCN \cdot nNH_3$ and $NiCl_2 \cdot 2NH_3$, the hydration has utilized $CaBr_2 \cdot 2H_2O$, the solvation has utilized TFE and E181, the solute mixture has utilized LiBr and $CaCl_2$ and the clathrate has utilized $R11 \cdot nH_2O$ or $R142b \cdot nH_2O$.

### 10.2.3. Methods of improving the performance of compression heat pumps.

The heat pump cycles of the high efficiency compressor-driven heat pumps being developed consist of (1) A non-azeotropic mixture working fluid multi-stage compression condensation cycle, and (2) A non-azeotropic mixture working fluid two-stage economizer cycle. Schematic diagrams of these pump cycles are shown in Fig. 10.2.5. Details are expained below.

The non-azeotropic mixture working fluid multi-stage compression condensation cycle of Fig.10.2.5-(1) contains a 50°C waste heat source which is used to produce 85°C hot water. This cycle is intended for air-conditioning and hot water supply. The heat source fluid enters the heat pump at 50°C and leaves it at 45°C. On the other hand, the heat load fluid is supposed to enter the heat pump at 50°C and leave it at 85°C. Therefore, the ideal cycle is expected to take the form indicated by dotted lines in the diagram. The maximum COP value estimated on the basis of ideal conditions of heat exchange at zero temperature difference will be 17, to be obtained from COP =1 / [1-{318K (heat source fluid outlet temperature) + 323K (heat source fluid inlet temperature) } / {323K (heat load fluid inlet temperature) + 358K (heat load fluid outlet temperature) } . Although the target COP to be actually developed is 8, two approaches are being considered with respect to the cycle, in order to make the cycle identical to that indicated by the dotted line. One is the usage of the 3-stage condensation cycle, and the other is the usage of a non-azeotropic mixture working fluid to realize the gradual change of temperature along the evaporation and condensation processes in the heat exchanger modifying the cycle into the Lorenz cycle. Briefly speaking, the effect of the 3-stage condensation cycle is that it will not require a 85°C temperature to heat the 50°C load fluid up to 60°C. When the working fluid is heated to 65°C by the first compressor, it is directed to the condenser, bypassing the next stage compressor to heat the load fluid. The cycle is thus modified to realize the Lorenz cycle by generating three temperature levels (e.g., 71°C, 80°C, 91°C) in one heat pump cycle. A schematic diagram of this multi-stage compression condensation cycle is shown in Fig.10.2.6-(1).

Furthermore, the development of heat exchangers for the phase change heat transfer of non-azeotropic mixtures has been an important problem. In the condenser of the non-azeotropic mixtures, the less condensable gas has been gathered near the heat transfer surface to be the resistance for diffusion and to make the boundary layer for mass transfer, and the condensation heat transfer coefficients have been decreased to one-tenth compared with those for one component. In this project, the shell & tube type condenser was innovatively revised to realize the large enhancement and to be economically feasible by setting the partition plate to prevent the diffusion along the streamline, by making the smaller cross section in the shell side, by increasing the speed of the vapor due to the subsequent subcooler and by extracting the less condensable gas gathered near the heat transfer surface to the indirect heat

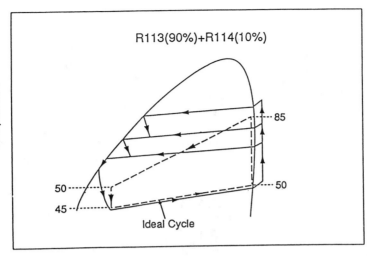

(1) non-azeotropic mixture working fluid multi-stage compression condensation cycle

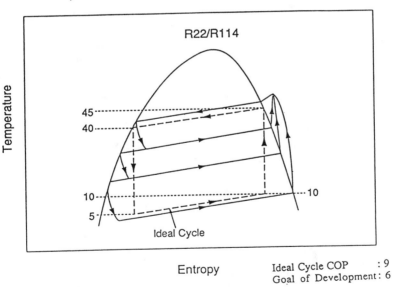

(2) non-azeotropic mixture working fluid two-stage economizer cycle

Fig. 10.2.5. Characteristics of heat pump cycles of high efficiency compression heat pump

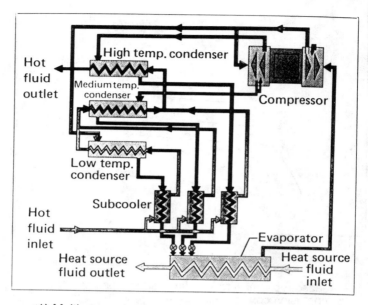

(1) Multi-stage condenser system flow diagram

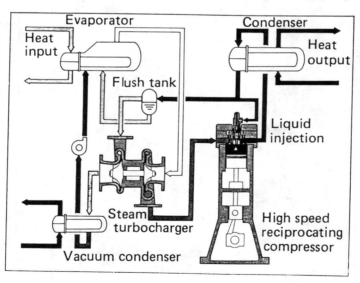

(2) High Temperature Heat Pump system flow diagram

Fig.10.2.6. Schematic diagrams of high performance heat pump systems [10.2.12]

exchanger in the evaporator. Since this heat pump could operate efficiently over a wide range of heat source temperatures, this heat pump could be applied to the lower temperature region where the larger heat demand exists for the hot water supply, such as in the case where the COP could be 8 for the heat load temperature rise from 35°C to 65°C, and also to the higher temperature region of about 100°C thermal output for industrial use.

The non-azeotropic mixture working fluid two-stage economizer cycle of Fig.10.2.5-(2) is intended for both the cooling and the heating related to the air conditioning. For the case of heating, the heat source fluid would enter at 10°C and come out at 45°C. The maximum expected value COP for the ideal cycle is 9.0, and the development goal is 6. Several approaches are being made to bring the actual cycle close to the ideal cycle.

One is the usage of a two-stage economizer, in which the economizer process is repeated twice so that the working fluid coming out from the condenser is flushed and evaporated, and only the vapor is returned to the middle point of the compressor for its re-compression. Briefly speaking, effects of this two-stage economizer are that if re-compression is possible at an intermediate stage before reaching evaporation pressure, the power required by the compressor for attaining the same high temperature will be decreased. The other approach is the usage of the non-azeotropic mixture working fluid cycle to realize the Lorenz cycle. The above-mentioned two kinds of high performance heat pumps will use environmentally acceptable alternative working fluids for their pilot plant as shown in Table 10.2.3 and as explained in Chapter 6.

The target COP values of the super heat pump project are about double those of the conventional levels. In order to achieve this high performance, the above mentioned improvements for the composition of heat pump cycles will not be sufficient. Many technological improvements, such as realizing high performance compressors ( improved shape of screw teeth and vanes, spray of liquid to make the compression power more efficient by realizing the compression process nearly along the vapor-liquid saturation line), high performance heat exchangers (reduction of temperature difference in heat exchangers), recovery of heat generated from compressor driving motor, etc., will enable reaching such a high target value. For example, the two-stage economizer will bring about 10% of improvement, non-azeotropic mixture working fluid will bring about 20%, high performance compressor 10%, recovery of heat of motor 5%, and high performance heat exchanger about 30%. Thus, COP of double the conventional level would be achieved when all of these technological improvements are successfully made. In addition to the usage of the above-mentioned multi-stage condensation system, two-stage economizer, and non-azeotropic mixture working fluid, this project also includes research studies of elemental technologies for the improvement of the cycle performance, such as a expansion turbine and subcooler. The composition of the cycle for achieving maximum performance will become more and more complex.

Secondly, research and development of approximately 150°C and 300°C high temperature thermal output heat pumps in the super heat pump projects have also been conducted. Schematic diagrams of their cycle characteristcs are shown in Fig.10.2.7. Details are explained below.

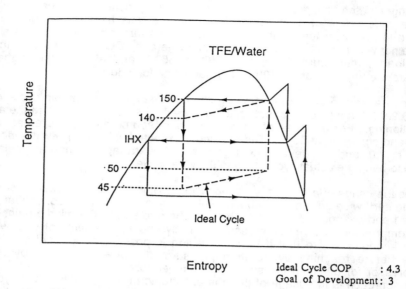

(1) Two-stage compression cycle with a direct contact intermediate heat exchanger

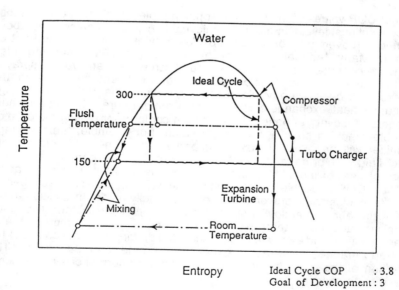

(2) Flush vapor recovery turbochanger assisted 300°C output cycle

Fig. 10.2.7. Cycle characteristics of high temperature type compression heat pump

The two-stage compression cycle with a direct contact intermediate heat exchanger (Fig.10.2.7-(1) ) utilizes a 50°C waste heat source to produce a high temperature of about 150°C for industrial heating applications. The heat source fluid is supposed to enter at 50°C and come out at 45°C, while the heat load fluid enters at 140°C and comes out at 150°C. The expected maximum COP value of this operation is 4.4, and the actual development target value is 3. In order to achieve the goal, a two-stage compression cycle and a direct contact type intermediate heat exchanger are used. Although the compression stage is limited to two-stage due to the restriction of compression ratio of the screw compressor, the direct contact intermediate heat exchanger is expected to bring the economizing effect obtained by intermediate cooling compression process and flush evaporation. In addition, a two-phase expansion turbine has been developed for the recovery of several percentages of the total power in place of the expansion valve. Concerning the compressor, the development of thrust offset structured rotors was conducted to realize higher volumetric compressor efficiency. Furthermore, the working fluids for the high temperature heat pump of about 150°C thermal output have been researched and selected, and the thermophysical properties have been measured for TFE, 4FP, 5FP and their azeotropic mixtures with water and also n-perfluorooctane. Since this heat pump could be applied for the wider range of the heat source temperature, COP of 5 will be achieved for the temperature rise from 95°C to 150°C.

The cycle with a single-stage flush vapor supercharger of Fig.10.2.7-(2) is supposed to use a 150°C heat source for producing a 300°C high temperature for industrial heating applications. One characteristic of this high temperature heat pump system is the usage of water as the working fluid and the water could be one of the promising working fluids for the higher temperature heat pumps. Inlet/outlet temperatures are supposed to be 150°C for the heat source side and 300°C for the heat load fluid side. Heat exchange is supposed to be done by phase change of the steam. The maximum expected COP value for this operation is 3.8, while the target value of actual development is 3. To achieve this goal, the heat cycle is devised such that flush equipment would be used for generating water vapor, which then would be adiabatically expanded to approximate room temperature for recovering power to drive the turbo-compressor. The vapor expanded to approximate room tempereture mixes with the high temperature condensed liquid in the flush equipment and returns to the evaporator at about 150°C, as shown in Fig.10.2.6-(2). As for the compression process, the steam was compressed first by the recovery turbo-compressor and then compressed to 90 atmosphere by the reciprocating compressor to make a thermal output of 300°C. This flush vapor turbo-compressor would have equivalent cycle characteristics with a two-phase expansion turbine. Furthermore, liquid water was injected in the compressor to prevent excessive superheat of the steam and to improve the cycle performance. The results of the fundamental research of the two-phase compression process are shown in Fig.10.2.8. As seen from the figures, the two-phase compression process has the ability of improving of COP by about 6%. [10.2.6] The mechanism of this improvement is based on the reduction of the excessive superheat by using the two-phase compression process to make the heat pump cycle similar to the ideal cycle of Carnot, as seen from the figure.

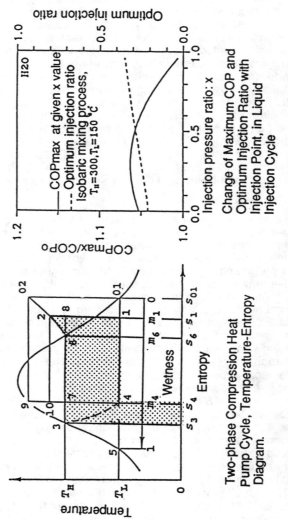

Fig. 10.2.8. Effects of two-phase compression process

In this heat pump cycle, some amount of the heat was transferred to the ambient cooling water, which was equivalent to the power recovery cycle between the temperature of the thermal output of about 300°C and the ambient temperature of about 25°C. Furthermore, in this cycle, some amount of the working fluid had its temperature decrease to the ambient level. By exchanging heat between the working fluid of the ambient temperature and the heat source steam of about 150°C, the heat source steam decreased its temperature greatly to have the temperature close to the ambient temperature. This means that the heat source fluid utilized not only its latent heat but also its sensible heat and that the amount of the exhausted heat to the environment by the disposal of the heat source fluid is also decreased. Concerning the turbocharger, heavy load equipment for the erosive working fluid of steam has been developed. Furthermore, this heat pump system can be applied to the wider range of the heat source temperature by changing only the component of the turbocharger and is estimated to be the COP of 6 for the temperature increase of 100K from 200°C to the thermal output of 300°C.

The technology for the important and innovative components has been also developed and evaluated especially for heat exchangers and the working fluids. For the evaporator of the non-azeotropic mixtures, the segmented flow channel reversal two-phase flow type has been developed to revise the performance of shell & tube evaporator. In this heat exchanger, the working fluid was first evaporated outside of the surface manufactured tube in the smaller dryness fraction region and then evaporated inside of the surface manufactured tube in the larger dryness fraction region. Therefore, the performance of this evaporator has achieved a heat transfer enhancement of 3.0 times and a pressure loss of 0.7 times as compared with those of smooth tubes. An EHD (electro-hydrodynamical) condenser has been developed newly in this project for the condenser of the high temperature heat pump. The performance of this EHD condenser achieved about 6 times the heat transfer coefficients for the same film Reynolds number compared with those of outside condensation of the vertical smooth tube. This EHD condenser would be effective for the longer vertical condenser and for the organic working fluid of the smaller latent heat with negligibly small electric power consumption. [10.2.7] Also, a high performance stainless plate-fin heat exchanger has been developed for the heat exchangers of the non-azeotropic mixtures and for the heat exchangers as a component of heat pumps and chemical storage systems. As alternatives for CFC working fluids in pilot plants, several environmentally-acceptable alternative unregulated organic working fluids such as R123, R134a and R142b have also been researched and evaluated, which has contributed to the world-first data book of thermophysical properties of environmentally acceptable fluorocarbons of HFC 134a and HCFC 123, as published by the Japanese Association of Refrigeration.

As described above, in the development of high performance compression heat pumps, all heat pump systems have achieved their high and difficult targets through the development of important breakthroughs and are expected to realize the final goals of the pilot systems of about 1000kW scale.

10.2.4. Contribution of "Super Heat Pump" to global environmental protection

The contribution of the high performance heat pump, such as the super heat pump, to energy conservation and global environmental protection, with their economical aspects, is investigated in this subsection.

Concerning the progress of energy conservation, by utilizing the high performance super heat pump, the energy conservation effects will be roughly estimated for the case of Japan as a typical example of an industrial country, though the quantitative estimation of the application field and the ratio of heat pumps would be difficult. In Table 10.2.4., the estimated energy consumption relating the application field of heat pumps in the years 1985 and 2000, and the estimated ratio of heat pump are shown. [10.2.11]. Among the various application fields of heat pumps, residential use, which contained mainly small scale distributed heat demand, is not included to the application field of the super heat pump. Concerning air conditioning of the buildings, district area and factories, one third of the estimated ratio of the heat pumps, which is 18%, is assumed to be occupied by the super heat pump in the year 2000. Concerning the hot water supply of commercial buildings, one third of the estimated ratio, which is 2%, is assumed to be occupied by the super heat pump as a replacement for boilers and conventional heat pumps. Also, concerning the process heat of industrial use, half of the estimated ratio, which is 2%, is assumed to be occupied by the super heat pump as a replacement for the boiler. To estimate the energy conservation effects of the super heat pump in the amount of the primary energy, the following values are assumed: thermal efficiency of the boiler, 85%; heating COP of the conventional heat pump, 3; that of the super heat pump, 6; cooling COP of the super heat pump, 7; COP of the conventional heat pump for hot water supply, 4; that of the super heat pump, 8 ; COP of the super heat pump for the industrial process heat, 5; efficiency of electric power generation, 38.1%.

heating: $124 \times 10^{12} \times \{ 0.09 \times (1/0.85 - 1/6/0.381) + 0.09 \times (1/3 - 1/6)/0.381\}$
$= 13.1 \times 10^{12}$ kcal/y

hot water supply: $112 \times 10^{12} \times \{0.01 \times (1/0.85 - 1/8/0.381) + 0.01 \times (1/4 - 1/8)/0.381\}$
$= 1.3 \times 10^{12}$

cooling: $38 \times 10^{12} \times 0.18 \times (1/3.5 - 1/7)/0.381$     $= 2.6 \times 10^{12}$

process heat: $975 \times 10^{12} \times 0.02 \times (1/0.85 - 1/5/0.381)$     $= 12.7 \times 10^{12}$

As calculated above, the total energy conservation will be summed up to $29.7 \times 10^{12}$ kcal/year and the amount of the heating oil about 3 million kl/year in the case of Japan. This amount is about 2% of the total energy consumption in the application fields relating to the heat pump and would be equivalent to the amount of small scale power generation which has been called " local energy " and which has contained smaller scale water power plants and power generation plants from waste incineration facilities. This means that energy conservation, by utilizing the super heat pump, would be expected as if a new energy resource could be created.

Concerning the reduction effect of the emission of exhaust $CO_2$, the reduction effects are considered to be composed of the energy conservation effect and effect of the shift of the energy resources from oil to electricity where LNG and the nuclear energy are the main energy resources. To calculate the reduction effect of $CO_2$ emission, the following assumptions are made; amount of $CO_2$ emission for electric power, 0.04g/kcal (primary energy basis); the amount of $CO_2$ emission from the boiler, 0.085g/kcal; the composition of the primary

Table.10.2.4.
Estimated energy consumption relating the application field of heat pumps in the year 2000 and the estimated ratio of heat pumps

| application field | use | amount of energy consumption (kcal/year) | | | estimated ratio of heat pumps | |
|---|---|---|---|---|---|---|
| | | 1985 | 2000 | average increasing rate per year (%) | 1985 (%) | 2000 (%) |
| residential | heating | $117 \times 10^{12}$ | $168 \times 10^{12}$ | 2.6 | 8 | $52 \pm 12$ |
| | hot water supply | $113 \times 10^{12}$ | $162 \times 10^{12}$ | 2.6 | 0 | $12 \pm 5$ |
| commercial | heating | $83 \times 10^{12}$ | $119 \times 10^{12}$ | 2.6 | 13 | $54 \pm 14$ |
| | hot water supply | $78 \times 10^{12}$ | $112 \times 10^{12}$ | 2.6 | 0 | $6 \pm 3$ |
| | cooling | $21 \times 10^{12}$ | $38 \times 10^{12}$ | 4.0 | 13 | $54 \pm 14$ |
| industrial | process heat | $861 \times 10^{12}$ | $975 \times 10^{12}$ | 0.8 | 0 | $3.6 \pm 1.3$ |
| | heating | $4 \times 10^{12}$ | $5 \times 10^{12}$ | 0.8 | 0 | $54 \pm 14$ |

___ Application field of Super Heat Pump

energy for electricity generation in the year 2000, 47% of nuclear energy, water power and geothermal energy, 11% of coal, 22% of oil, 20% of LNG; the amount of $CO_2$ emission in the weight of carbon, 0.098g/kcal of coal, 0.085g/kcal of oil, 0.056g/kcal of LNG.

heating: $124 \times 10^{12} \times \{0.09 \times (1/0.85 \times 0.085 - 1/6/0.381 \times 0.041) + 0.09 \times (1/3-1/6)/0.381 \times 0.041\}$ = $1.1 \times 10^{12}$ g/ycar

hot water supply: $112 \times 10^{12} \times \{0.01 \times (1/0.85 \times 0.085 - 1/8/0.381 \times 0.041) + 0.01 \times (1/4-1/8)/0.381 \times 0.041\}$ = $0.1 \times 10^{12}$ g/ycar

cooling: $38 \times 10^{12} \times 0.18 \times (1/3.5 - 1/7)/0.381 \times 0.041$ = $0.1 \times 10^{12}$ g/year

process heat: $975 \times 10^{12} \times 0.02 \times (1/0.85 \times 0.085 - 1/5/0.381 \times 0.041)$
= $1.5 \times 10^{12}$ g/ycar

From the above calculation, the total reduction amount of $CO_2$ emission will be summed up to $2.8 \times 10^{12}$ g/year, which is about 2.8 million tons per year in the weight of the carbon and which is equal to about 1% of the total emission of $CO_2$ in the case of Japan.

Concerning the economical aspects of the super heat pump energy accumulation system, it would be difficult to estimate the initial cost of the super heat pump, since super heat pumps are still under development. Therefore, the operating cost estimated from the target performance of energy conversion will be calculated for the cooling condition of the following four cases.

Conventional HP(COP=3.5) : $54/Gcal
Conventional HP + Thermal storage (ice) : $42/Gcal
Super Heat Pump(COP=7) : $27/Gcal
Super Heat Pump + Chemical storage : $17/Gcal

To calculate the operating energy cost, in the case of Japan the following conditions are assumed: the COP of the conventional heat pump for the cooling condition, 3.5; the COP of the conventional heat pump with the usage of the thermal storage system of ice, 2.7; the efficiency of the chemical storage, 75%; the cooling period for one day, 10 hours (from 8hr to 18hr); monthly operating time, 25.5 days; the supply of half of the cooling load by the heat pump in a day's time; the currency 145 yen/$; base cost of electricity $10/kW; the additional cost of electricity $120/MWh; the additional cost of electricity for thermal storage system, $31/MWh; the efficiency of the thermal storage system of ice, 90%.

Furthermore, one new program of the utilization of "Urban Energy" has just been started by the Japanese Ministry of International Trade and Industry. This "Urban Energy" means the utilization of the exhaust heat of subways, buildings and incineration facilities of waste as the heat source of the super heat pump. Also, river water or the waste water from residential use and from factories are to be used as the heat source of the high performance heat pump. The above kinds of urban energy have not yet been widely utilized. Since there would be a temperature difference between the river water or the waste water and the atmospheric air, in the case of utilizing the urban energy as the heat source of the heat pump for the heating condition, the temperature increase necessary for the heat pump could be decreased and the COP of the heat pump will be raised. For example, by changing the heat source from the atmospheric air to the river water, the COP was assumed to be increased to 4.9 from 2.6. The trial calculation of the energy conservation effects, the reduction amount of the $CO_2$ emission and the reduction amount of $NO_X$ emission has been made for the typical examples of business area (Marunouchi, Tokyo, Japan, 14.6 ha), the commercial area (Ginza) and the residential area( Hikarigaoka). In the case of the business area, by utilizing the exhaust heat of subways and buildings, river water and the waste water from the buildings as the heat source of the super heat pump, the energy conservation effects will be 27%, the reduction ratio of NOx emission will be 69% and the reduction ratio of $CO_2$ emission will be 48%. To calculate the effects, the following conditions are assumed: the total floor area, 0.49 million m$^2$; the heat demand of the cooling, $3.7 \times 10^{10}$ kcal/y; the heat demand of the heating, $1.7 \times 10^{10}$ kcal/y; the exhaust heat of the buildings, $3.7 \times 10^{10}$ kcal/y; the exhaust heat of the subways, $1.4 \times 10^{10}$ kcal/y; the dissipation heat of underground power-transmission cables, $0.4 \times 10^{10}$ kcal/y; the exhaust heat of transformer substations, $0.1 \times 10^{10}$ kcal/y; the extracted heat from the river water, $493 \times 10^{10}$ kcal/y; the extracted heat from the waste water of the sewage system, $136 \times 10^{10}$ kcal/y; the replacement from the conventional heat pumps of the heat source of air and the boiler to the super heat pump utilizing the urban energy. Similar effects have been also calculated for commercial and residential areas: energy conservation, 28% and 36%; reduction of NOx, 59% and 76%; reduction of $CO_2$ emission, 42% and 58%, respectively.

From the above consideration, it has been clarified that the energy conservation effects and the reduction of $CO_2$ and NOx emission realized by the development of high performance heat pumps would be remarkable and that the necessity and the importance of the development of various kinds of higher performance heat pumps and various kinds of urban energy as the heat source of heat pumps would be seriously considered.

## 10.2.5. Concluding remarks

This paper presented an outline and characteristics of the "Research and Development of Super Heat Pump Energy Accumulation System" project and its expected contribution to the global environmental protection, emphasizing the high performance compressor-driven heat pumps and their newly developed elementary technologies. As explained in the results of the interim evaluations, the bench plants of the 100kW class have been successfully researched and developed.

At present, according to the increasing needs of energy conservation and the needs for the environmental protection technologies, this project has been actively promoted toward the targets of operation and research of 1,000kW class pilot systems to be completed by 1993, and the development of the higher performance heat pump systems has come under serious consideration.

## ACKNOWLEDGEMENTS

The author wants to express his sincere gratitude to the Moonlight Project Promotion Office of AIST, MITI, New Energy and Industrial Technology Development Organization (NEDO), Technology Research Association of Super Heat Pump Energy Accumulation System, National Chemical Laboratory for Industry, Prof. Ichiro Tanasawa of the University of Tokyo for his wonderful advice, and also to Dr. I. Yamashita, Mr. K. Ozaki, Mr. N. Endo, Dr. T. Munakata, Dr. M. Akai and other colleagues of the Mechanical Engineering Laboratory for their fruitful collaboration.

REFERENCE

[10.2.1] NEDO, The development of super heat pump energy accumulation systems, January 1990.

[10.2.2] NEDO, The development of super heat pump energy accumulation system, July 1990. No.1

[10.2.3] A. Fujii, Current status of the super heat pump energy accumulation system in NEDO projects, Proc. 3rd IEA Heat Pump Conf., Tokyo, 1990, p.349.

[10.2.4] M. Akai, A. Yabe, K. Ozaki, N. Endo and T. Munakata, Qualification of high performance heat pump--evaluation of COP for "Super Heat Pump" --, Proc. 3rd IEA Heat Pump Conf., Tokyo, 1990, p.787.

[10.2.5] JETRO (Japan External Trade Organization), Heat pump R&D in Japan, New Technology Japan, 1989.

[10.2.6] K.Ozaki, N. Endo, A. Yabe and K.Kobayashi, Basic study on high performance heat pump systems accompanying two-phase compression process, Proc. 1990 Int. Compressor Engineering Conf. Purdue U.S.A., p.183.

[10.2.7] A. Yabe, Y. Mori, and K. Hijikata, Heat transfer enhancement techniques utilizing electric fields, Heat Transfer in High Tech, and Power Engng., Hemisphere, 1987, p.294.

[10.2.8] H. H. Kruse, Current status and future potential of non-azeotropic mixed refrigerants, Proc. 1987 IEA Heat Pump Conf., Orlando U.S.A., Orlando U.S.A., p.173

[10.2.9] Asahi Flon, Technical report on mixture working fluids for thermal cycle (in Japanese)

[10.2.10] Sumitomo Precision CO.; SMALEX catalog.

[10.2.11] Heat Pump Technology Center of Japan, Survey of the spread of heat pump technology in Japan (in Japanese), 1988-7.

[10.2.12] Technology Research Association of Super Heat Pump Energy Accumulation system, Super Heat Pump Energy Accumulation system.

# 10.3 COGENERATION
## --- ITS EFFECTS AND PROBLEMS---

**Akira Yabe**
Mechanical Engineering Laboratory
Agency of Industrial Science and Technology
Ministry of International Trade and Industry
Tsukuba Science City, Ibaraki 305, Japan


10.3.1. Characteristics and effectiveness of cogeneration

The word "Cogeneration" means the simultaneous generation of power and heat, or the electric power and the heat, from one or several kinds of primary energy by the same equipment. This kind of equipment has been installed for over fifty years and utilized as independent power generation equipment for industrial use. In recent years, from the viewpoint of energy utilization, the characteristics of simultaneous generation of electricity and heat have attracted much attention and cogeneration systems have rapidly prevailed, mainly for residential use. The application cases are hotels, sports centers, office buildings, etc. In Europe, "Cogeneration" has been called "Combined Heat and Power Generation (CHP)".

The principle of cogeneration is to extract power by utilizing the heat of a given temperature first and then to use the heat itself. If we would make the heat transfer in heat exchangers without making any power recovery, this process would be equivalent to the heat loss process to the lower temperature heat source accompanied with the thermodynamic cycle of the power recovery. This means that the heat transfer without making any power recovery could be understood only as the loss process from the viewpoint of the available energy or the energy utilization. Therefore, according to the principle of the cogeneration, wherever the heat exists, to make the power recovery by utilizing the temperature difference between the heat and the atmosphere should be the first and most important process. The rationale of this method is to utilize the heat in series by making power recovery first and then by using the heat from the same equipment. This would be completely different from the conventional method of independent and parallel supply of electricity and heat. The principle of cogeneration is also equivalent to the principle of the multi-purpose utilization or the cascading utilization of heat.

To evaluate the two types of energy produced simultaneously by the cogeneration system would depend on the characteristics of the application cases, especially on the difference of the quality between the electricity and the heat. However, the effectiveness of the cogeneration system, which would be one of the most effective methods of energy utilization, could be explained by the application example of an engine-driven heat pump system. In an engine-driven heat pump system, where the power is generated by the engine, this power is directly utilized for the compressor power of the heat pump system and where the thermal output and the exhaust gas of the engine would be utilized for residential heating, the total thermal output will come to 1.64. This assumes the thermal efficiency of the engine to be 35% and the COP of the heat pump cycle to be 3 and the recovery ratio of the exhaust gas to be 90%, as derived by $0.35 \times 3 + (1-0.35) \times 0.90 = 1.64$. Therefore, by as compared with the heating value of fossil fuel, over 1.5 times the thermal output could be drawn and to use this kind of cogeneration system would be more effective for the heating demand and for the hot water supply, as compared with the boiler system. Furthermore, the important and additional characteristics of the system are to be adopted into various kinds of higher performance heat pump systems in order to realize the larger thermal output, as compared with the heat pump system itself. For example, by adopting the higher performance heat pump system of its COP of 6 into the cogeneration system, over 2.5 times the thermal output as derived by $0.35 \times 6 + (1-0.35) \times 0.90 = 2.69$ will be generated, as compared with the boiler system. This applicability to the future higher performance heat pump systems and to the higher performance thermal cycle would be an important characteristic denoting the effectiveness of the cogeneration system with the utilization of the exhaust gas of the engine. Two examples of the cogeneration systems are shown in Fig. 10.3.1.

10.3.2. Current status and research & development of cogeneration.

Cogeneration systems may be generally divided into two types: large scale centralized cogeneration systems for industrial use and small scale distributed cogeneration systems for residential and commercial uses. The large scale cogeneration systems for industrial use have been operated for over fifty years in Japan for a simultaneous supply of steam and electricity. The industrial cogeneration system has been mainly composed of the boiler and the steam turbine cycles. The industrial cogeneration systems of a scale larger than 500kW had the ability of generating electricity of 15.5 million kW (2.7% increase compared with the previous year) in the year 1987 (for Japan), which was 8.9% the total electric power generation ability, and the total amount of generated electricity had attained 74.8 billion kWh (10% increase compared with the previous year) in the year 1986, which was 11% of the total electric power generation. In the cases where the balance of the output of electricity to heat would be adequate to the necessary balance of the applications, the cogeneration system has performed positively in, for example film factories, food processing companies, etc. The internal combustion engines for industrial cogeneration were composed of gas turbines, which had 370MW for 39 cases, Diesel engines, which had 300 MW for 88 cases and gas engines which had 30 MW for 50 cases in the year 1988.

Concerning the small scale distributed cogeneration for residential and commercial use, several kinds of small but efficient types of engines have been utilized or under development. Promising equipment for generating power and electricity are Diesel engines, gas engines and several kinds of systems under development e.g., gas turbines, Stirling engines and fuel cells. In the case of Japan, Diesel engines are 70 MW of 119 cases, gas engines are 30 MW of 112 cases and gas turbines are 10 MW of 8 cases. For residential and commercial uses, there are many cases, which have a relatively large ratio of heat demand for the hot water supply, e.g. hotels, office buildings, sports facilities, research laboratories, training facilities, hospitals, super markets,

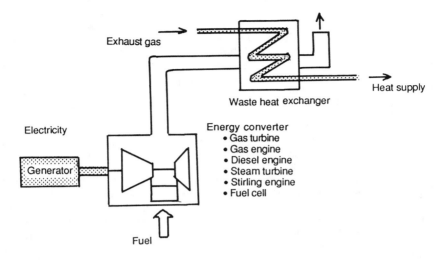

(1) Electricity and heat supply

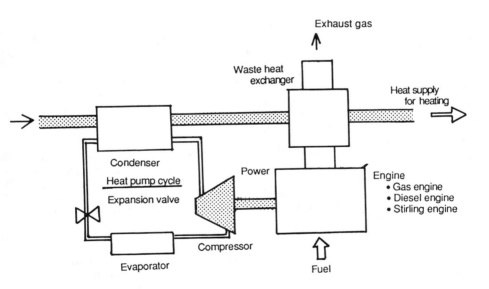

(2) Power and heat for heat supply

Fig.10.3.1. Two examples of cogeneration system

schools, computer centers and district air conditioning facilities. To promote its further spread and to make wider application fields, the key points are looking for the uses having a larger amount of heat demand and realizing the district heating at many places. Concerning the amount of output of electricity for one facility, the average is 600 kW for hotels, which occupy about half of the total demand of electricity. In planning residential and commercial facilities, to grasp the pattern of the heat demand correctly, which would have hourly and large fluctuations through annual usage, is important in making the optimum cogeneration system.

Concerning thermal efficiency to generate electricity, to raise the temperature of the combustion gas would bring greater efficiency. In the "Ceramic Gas Turbine (CGT) Project" conducted by the Japanese Ministry of International Trade and Industry, 300 kW ceramic gas turbines for industrial cogeneration are being researched and developed for 9 years, starting from 1988. The goals of the development are an inlet temperature of the turbine of 1350°C and a thermal efficiency of the engines of over 42%. Also, the compared conventional metal blade small scale gas turbines have had an inlet temperature of about 900°C and a thermal efficiency of 15~20%. By realizing this ceramic gas turbine, the efficiency of generating electricity will be increased with the combination of the steam turbines utilizing the exhaust gas of the ceramic gas turbine. This technology would be an important technology for the development of cogeneration and energy conservation.

The total thermal efficiency of generating electricity for the large scale power plants has already increased over 43% by utilizing the top stage of an advanced gas turbine. The thermal efficiency of the ceramic gas turbine will not go ahead of that of the large scale advanced gas turbine by the scale effect, but by considering the dissipation loss of cables for the electricity transmission and the advantage of the widely distributed small scale systems, which are near the places demanding the electricity, a thermal efficiency of 30~40% is considered to be equivalent to that of the large scale electric power station.

Concerning the fuel cell which has been also under development by the Japanese energy conservation project nicknamed "Moonlight Project", the goals are an electric power generation efficiency of 37~40% for a 200kW cogeneration system. In this fuel cell system, high temperature exhaust gas of over 150°C could be utilized for the heat demand. Since a difficult problem in the maintenance of the cogeneration system is obtaining a technician for the operation and the maintenance, the development of a cogeneration system of high reliability which would have the maintenance-free period of over 100,000 hours is important.

A difficult problem of cogeneration systems is matching the demand of electricity and thermal outputs and the limited flexible range of changing the ratio of the output of electricity to the thermal output. This has been one of the barriers to wider usage. But, to remove these shortcomings, the development of the "Chain Cycle" has been conducted in order to make a wider flexible range for changing the ratio of the output of electricity to the thermal output. The structure of the chain cycle is shown in Fig. 10.3.2 and is composed of the gas turbine cycle and the steam turbine cycle. For increasing the ratio of the output of electricity, the steam generated by the exhaust gas of the gas turbine is injected into the combustor of the gas turbine to increase the flow rate of the

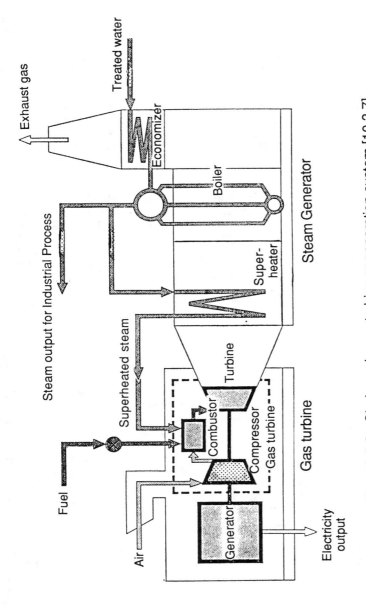

Fig. 10.3.2. Chain cycle gas turbine cogeneration system [10.3.7]

turbine and to increase the generating electric power. One developed facility has the ability of changing the ratio of the output by the electric power by 0~2,300 kW and the amount of the generated steam by 0~8 t/h.

There is a connection problem with conventional electric cables, since the surplus of the electricity output to the demand frequently occurs for the production of electricity in the cogeneration systems. In the case of Japan, investigations for making the new laws are being made following official technical guidelines. But, there is not any obligation for the electric power companies to take over the surplus of the electric output generated in the cogeneration systems. In the United States of America however, there is a law named "Public Utility Regulatory Policy Act of 1978" (abbreviatied: PURPA). According to this law, electric companies have to purchase the surplus of the electricity generated by the cogeneration plants for a reasonable price and supply the necessary amount of reserve electricity to the cogeneration plants for a fair and reasonable price. This law has promoted the wide spread of cogeneration systems. Therefore, In the U.S.A., cogeneration facilities, which were approved in March,1986, have increased to over 900 locations and their total amount of electric power generation has exceeded about 2.7 million kW. This includes the industry cogenerations such as chemical engineering plants, pulp and paper companies, petroleum refining factories and steel companies, and the cogeneration equipment for commercial and residential uses, such as hospitals, schools and commercial buildings. Similar situations of supporting systems of the spread of the cigeneration plants have been observed in several countries, i.e., United Kingdom and Italy.

10.3.3. Environmental aspects of cogeneration and problems to be solved

The energy conservation effects of cogeneration could be remarkable, as described in subsection 10.3.1. In particular, for a system with only a heat demand, the cogeneration system such as the gas engine heat pump, can produce larger energy conservation effects, as compared to those of high performance heat pumps. However, in many cases where the combination of electric power and the thermal output is utilized in facilities, some standards of evaluating the cogeneration systems have to be established. Since there is much difference in the quality of electricity or power and heat, the total thermal efficiency determined by the addition of the efficiency of electricity generation and thermal efficiency for providing the heat would not be adequate and additional evaluating methods would have to be established based on the viewpoint of available energy or based on the economically weighted addition of the electricity and the heat. The establishment of adequate evaluation methods are especially important for comparison with high performance heat pump systems, and for comparison among the various kinds of cogeneration systems at given temperature conditions.

Concerning the environmental aspects of the cogeneration systems, many cogeneration systems have to face the severe problem of regulations of exhaust gases. Due to the present technological conditions of the internal combustion engines, cogeneration systems with using internal combustion engines have not yet cleared the regulation standards of exhaust amount of NOx in Japan. The present condition of NOx amount of the exhausted gas for the engines of cogeneration systems at a thermal output of 200kW class in the year 1985, the gas engines had NOx of 3,300ppm, the Diesel engines had

NOx of 2,700ppm and the gas turbines had NOx of 540ppm, in the case where the concentration of the oxygen was assumed to be 0%. However, the guidelines of the reduction of NOx in metropolitan Tokyo, the capital city of Japan, indicates that the gas engines have to exhaust NOx below 300ppm, the Diesel engines have to exhaust NOx below 500ppm and the gas turbines have to exhaust below 200ppm. Furthermore, this guideline will be stricter from 1992, in that gas engines have to exhaust below 200ppm, Diesel engines have to exhaust below 300ppm and gas turbines have to exhaust below 150ppm. Compared with the Japanese national emission standards, where Diesel engines have to exhaust below 2,494ppm and where gas turbines have to exhaust below 236ppm (in the case where the concentration of the oxygen is assumed to be 0%), the standards of Tokyo are very severe, especially below one fifth of the NOx emission national standards/ for the Diesel engines and much smaller than the present conditions of the NOx emission of the actual engines. According to the standards of the exhaust NOx for the stationary internal engines of Tokyo, the goals of the exhaust NOx will be able to be completed by the combination of fuel-lean combustion and the three-way catalyst for gas engines, by the combination of the delay of the fuel injection process and the exhaust gas denitrification for Diesel engines and by the combination of steam injection and exhaust gas denitrification for gas turbines. However, the exhaust gas denitrification of the non-catalytic method, by use of ammonia ($NH_3$-SCR method), which has been frequently utilized for large electric power stations, has economical problems when applied to smaller cogeneration plants due to the necessity of making larger facilities for ammonia recovery. Furthermore, the three-way catalyst for the exhaust gas denitrification has been shown to be ineffective for the fuel-lean combustion conditions in the Diesel engines. Therefore, it is understood that the small scale internal combustion engines have many problems to be solved in their methods of exhaust gas denitrification, which is a large barrier in realizing larger usage and wider applications. Similar severe emission standards are in practice also in Kanagawa and Osaka Pref. of Japan and will be spread to other areas. The details of the reduction methods of NOx are explained in chapter 7.

As for the external engine of Stirling engines which have relatively clean exhaust gas, and the fuel cell, the problem of exhaust gas will be small enough and they will be widely used in the case that the cogeneration facilities are economically feasible.

Concerning the reduction of the emission of $CO_2$, the cogeneration systems have much advantage in reducing the necessary amount of fossil fuel consumption. However, compared with large scale facilities such as electric power stations, which have the future possibility of reducing $CO_2$ emissions by the separation of the carbon dioxide and by the discharge into the deep sea as explained in chapter 4, for the small scale stationary combustion engines the separation of carbon dioxide from the combustion gas and discharge into the sea would be difficult from an economical point of view. The reduction of the carbon dioxide emission from the engines themselves would therefore be difficult.

As explained above, the principle of cogeneration has the ability of realizing a larger contribution to the utilization of energy and the cogeneration systems with their exhaust gas emission standards satisfied, have rapidly spread to facilities which have an appropriate ratio of electricity demand to heat demand. For the Diesel and gas engines as the energy converter of the cogeneration systems however, there still exists a severe problem in clearing the emission standards and therefore it would be difficult for those engines to be widely used at the present technological level. Although the problems of NOx and $CO_2$ emission of exhaust gas would be carried to the medium and small scale stationary combustion engine facilities, desired technological breakthroughs will be obtained through future research and development for realizing the cascading usage of thermal energy and promoting energy utilization.

REFERENCES

[10.3.1] K. Hayakawa, S. Nakane ed., Cogeneration Handbook, (in Japanese), inoue-shoin, 1989.

[10.3.2] Cogeneration Research Society of Japan, Present Status of Non-industrial Cogeneration Systems in Japan, Cogeneration Research Society of Japan, 1988.

[10.3.3] Reports on the Optimization of the Cogeneration Systems, Activation Center of Petroleum Industries, 1988.

[10.3.4] T. Yoshii, Heat Pumps in Japan (2nd Edition), 8.2, Heat Pump Technology Center of Japan, 1988.

[10.3.5] Tokyo, Instruction Standards of NOx Emission from Stationary Internal Combustion Engines.

[10.3.6] M.Hirata, Present Status of Gas Turbine Cogeneration Systems and Measures to Environment Pollution, Proc. of 10th Anniversary Conference of "Moonlight Project", 1989.

[10.3.7] Present Status and Future Problems of Cogeneration (in Japanese), Energy, Vol.23-2,1990, pp.42-69.

[10.3.8] Y. Yamada, Research and Development of Ceramic Gas Turbines, Journal of Japan Society of Mechanical Engineers, Vol.93, No.855, 1990, p.129.

[10.3.9] Y. Tsutsui, etc., The Status of Ceramic Gas Turbine Programs in Japan, Proc. 28th Automotive Technology Development Contractors' Coordination Meeting, SAE, Dearborn, Michigan,1990.

[10.3.10] I. Yamashita, Practical Application of Stirling Engine is not too long, Journal of J.S.M.E., Vol.93, No.855, 1990.

# 10.4 HIGHLY DEVELOPED ENERGY UTILIZATION BY USE OF CHEMICAL HEAT PUMP

**Masanobu Hasatani**
Department of Chemical Engineering
Nagoya University
Furo-cho, Chikusa-ku, Nagoya 464-01, Japan


Introduction

Recently, it is urgently required for us to cope with the present energy problem which involves the countermeasure for the global contamination of environment as well as the development of energy saving and energy utilization technologies. As one of the solutions to the above-mentioned issue, chemical heat pumps are considered as one of the promising devices to meet with such a requirement. The reason for this is that the chemical heat pump in itself is the energy producing machine driven practically by the thermal energy; the chemical heat pump can operate with the combination of thermal energy of different temperatures with the minimum aid of mechanical energy, and it can reproduce more profitable thermal energy of which the temperature level is higher than the original one and/or lower than that of the environment.

Therefore, the chemical heat pump will make it possible to convert various heat sources like industrial waste heat, solar heat, terrestrial heat, etc. into more valuable heat sources. In addition, the chemical heat pump is expected to be applied to the load leveling of power stations by storing the electrical energy in the form of chemical energy, since the chemical heat pump naturally possesses the heat storing capacity. The chemical heat pump is also considered to display its potential power when combined with the cogeneration system by raising the total thermal efficiency of such an energy producing system.

In accordance with this background, it is considerd that the chemical heat pump may bring about an improvement of the total energy utilization efficiency relevant to the combustion of the primary energy sources like coal and petroleum. As a result, this improvement of energy utilization by use of chemical heat pump may bring about the diminution of emission of $CO_2$ which leads to the greenhouse effect. Furthermore, the chemical heat pump will be released from the flon problem which causes the depletion of the ozone layer by choosing appropriate chemical substances among many reaction candidates except flon derivatives.

In the following consecutive chapters, the operation principle and the fundamental characteristics of the chemical heat pump will be described by introducing the present research topics in the author's laboratory. The remaining problems to be solved will also be mentioned.

### 10.4.1 Example of Reaction Candidates for Chemical Heat Pump

Table 10.4.1 shows the typical examples of chemical reaction which are deemed as the candidates of constituent reaction for chemical heat pump[10.4.1]. It is obvious that the chemical heat pump is, in principle, composed of a reaction pair which works at a different temperature. The left and the right columns in Table 10.4.1 give the element reactions which take place at a higher and a lower temperatures, respectively. These constituent reactions are classified according to such operating reactant gas as $H_2O$, $H_2$, $NH_3$, $CO_2$, $CH_3NH_2$ and others. Table 10.4.2 shows in more detail the examples of reaction pair now under investigation for the purpose of applying them to absorption, adsorption, metalhydride heat pumps, and chemical heat pumps which are driven by the chemical reaction between inorganic salts and $H_2O$, $NH_3$, and so on.

### 10.4.2 Chemical Heat Pump by Use of $CaO/H_2O/Ca(OH)_2$ Reaction System

Among the reaction systems shown in Tables 10.4.1 and 10.4.2, $CaO/H_2O/Ca(OH)_2$ reaction system is considered as one of the promising reaction candidates for a high-temperature chemical heat pump. The enthalpy change between the exothermic hydration of CaO and the endothermic dehydration of $Ca(OH)_2$ is shown in Figure 10.4.1[10.4.2]. The distinctive features of this reaction system are; i) high density of thermal energy storage, over 1250kJ/kg, ii) high output temperature, over 673∼773K, iii) high reaction rate, iv) good reversibility of the reaction, v) good repetition of the reaction, vi) no toxicity and corrosion resistance of the reactant, vii) abundance of CaO resources and low cost to obtain the reactant, and so on. The items i) and ii) satisfy the goal of Moon-Light Project for developing the Super Heat Pump Accumulation System, that is, the heat storage capacity; more than 210kJ/kg, the output temperature; over ∼473K. On the other hand, the problems to be solved are; i) the pulverization of the reactant caused by the repetition of reaction cycle, ii) carbonization of CaO, iii) relatively an elevated temperature for the regeneration of CaO, more than 573K.

In view of the above-mentioned matter, the present article investigates both the heat transfer and reaction behaviours of the reactant particle bed in a manufactured lab-scale experimental heat pump unit. In this heat pump apparatus, the thermochemical reaction of $CaO/Ca(OH)_2$ along with the evaporation/condensation of water take place in the reactor and the evaporator/codenser, respectively. Based on the basic results obtained from the experiments, the applicability of this reaction system to a high-temperature chemical heat pump will be examined.

### 10.4.3 Operation Principle of $CaO/H_2O/Ca(OH)_2$ Chemical Heat Pump

Table 10.4.1 Examples of the reaction pairs for chemical heat pumps

| High-temperature reaction | Low-temperature reaction |
|---|---|
| $MgCl_2 4H_2O(s) = MgCl_2 2H_2O(s) + 2H_2O(g)$ | $H_2O(g) = H_2O(l)$ |
| $Ca(OH)_2(s) = CaO(s) + H_2O(g)$ | $H_2O(g) = H_2O(l)$ |
| $NiCl_2 6NH_3(s) = NiCl_2 2NH_3(s) + 4NH_3(g)$ | $CaCl_2 8NH_3(s) = CaCl_2 4NH_3(s) + 4NH_3(g)$ |
| $CaCl_2 8NH_3(s) = CaCl_2 2NH_3(s) + 6NH_3(g)$ | $NH_3(g) = NH_3(l)$ |
| $CaCl_2 6CH_3NH_2(s) = CaCl_2 2CH_3NH_2(s) + 4CH_3NH_2(g)$ | $CH_3NH_2(g) = CH_3NH_2(l)$ |
| $CaCl_2 2CH_3OH(s) = CaCl_2(s) + 2CH_3OH(g)$ | $CH_3OH(g) = CH_3OH(l)$ |
| $CaNi_5H_4(s) = CaNi_5(s) + 2H_2(g)$ | $LaNi_5H_6(s) = LaNi_5(s) + 3H_2(g)$ |
| $LaNi_5H_6(s) = LaNi_5(s) + 3H_2(g)$ | $MmNi_5H_6(s) = MmNi_5(s) + 3H_2(g)$ |

Table 10.4.2 Examples of the reaction system for chemical heat pumps

| | |
|---|---|
| Liquid/ Liquid(Gas) system | $NaOH/H_2O$, $H_2SO_4/H_2O$, $LiBr/H_2O$, R-11, $12/H_2O$, $NaSCN/NH_3$, $NaI/NH_3$, $NH_3/H_2O$, $LiBr/CH_3OH$, $LiBr$-$ZnBr_2/CH_3OH$, $C_6H_{12}/C_6H_6/H_2$, $(CH_3)_2CHOH/(CH_3)_2CO/H_2$, TFE/E181, $SO_2/O_2/SO_3$, etc. |
| Gas/Solid system | Zeolite/$H_2O$, Silica gel/$H_2O$, Activated carbon/$H_2O$, $Na_2S/H_2O$, $CaO/H_2O$, $MgO/H_2O$, $CaSO_4/H_2O$, Metal hydride/$H_2$, $NiCl_2/NH_3$, $CaCl_2/NH_3$, $CaCl_2/CH_3OH$, etc. |

The proposed chemical heat pump consists of the chemical reaction of the hydration/dehydration of $CaO/Ca(OH)_2$ and the evaporation/condensation of water as given in the following equations:

$$CaO(s) + H_2O(g) \rightleftharpoons Ca(OH)_2(s) + 1.858 \times 10^3 \text{ kJ/kg} \quad (10.4.1)$$

$$H_2O(g) \rightleftharpoons H_2O(l) + 2.316 \times 10^3 \text{ kJ/kg} \quad (10.4.2)$$

Figure 10.4.2 shows the relationship between the reaction equilibrium pressure Pe and temperature for the reaction in Equation 10.4.1. This is given by line (2)-(4) in the figure. Also shown is the relationship between the saturated steam pressure Ps and temperature for Equation 10.4.2. This is given by line (1)-(3). The principle of operation of this chemical heat pump is explained for the heat-release and heat-storing modes with the relation shown in Figure 10.4.2.

(i) <u>Heat release(temperature upgrading) mode</u>
As shown in Figure 10.4.3, consider a hermetically sealed reaction system having a reactor and an evaporator filled with CaO and water, respectively. If heat ($Q_M$) is added to the evaporator from a medium temperature ($T_M$) heat source, the water in the evaporator becomes pressurized steam (path 1-3 in Figure 10.4.2). Due to the pressure difference between Ps and Pe, this steam enters the reactor to undergo an exothermic hydration reaction with CaO. This causes the temperature in the reactor to rise (2→4) to temperature $T_H$ at which point the pressure, Pe is equal to Ps and the high temperature heat ($Q_H$) becomes available. According to the relation shown in Figure 10.4.3, the heat of which the temperature $T_H$ is 1205K can be, in principle, reproduced finally by using steam under subcritical condition ($T_M$=641K, Ps=27.5 MPa).

(ii) <u>Heat storage (regeneration) mode</u>
The $Ca(OH)_2$ produced in the heat release mode is regenerated to CaO by the following procedure. In this case, heat $Q_M$ from a medium temperature ($T_M$) source is added to the reactor which contains $Ca(OH)_2$ formed as described above. At the same time, the condenser is cooled to a temperature $T_L$. Under these conditions, the $Ca(OH)_2$ undergoes an endothermic dehydration to regenerate CaO by releasing steam. The steam released in the reactor moves to the condenser(2→1 in Figure 10.4.2) due to the pressure difference between the two chambers. There it condenses by releasing its latent heat of condensation to the low temperarure heat sink($T_L$). So long as the regenerated CaO is separated from water by detaching the reactor from the condenser, the reaction heat can be stored for a long time in the form of chemical energy.

10.4.4 Chemical Heat Pump Experiment

Photo 10.4.1 shows the appearance of chemical heat pump unit employed in the author's research group [10.4.3]. Also shown in photo 10.4.2 is the inside of the reactor. Figures 10.4.4 and 10.4.5 show a schematic diagram of the experimental unit and the detail of the reactor, respectively. The heat pump unit consists of the reactor and the evaporator/condenser which are made of stainless steel(SUS304) and are cylindrical in shape, having both an inside diameter and a height of 150mm (the substantial inner volume is $2.4 \times 10^{-3} m^3$).

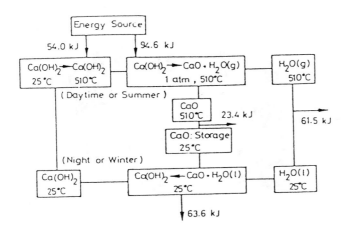

Figure 10.4.1 Enthalpy change of $Ca(OH)_2/CaO$ reversible reaction

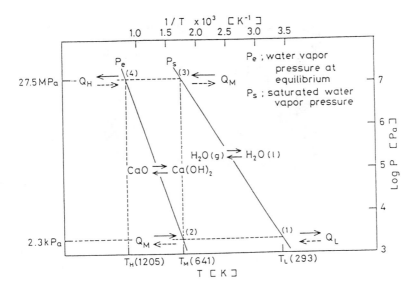

Figure 10.4.2 Temperature-Pressure line

Photo 10.4.1 Manufactured chemical heat pump apparatus

Photo 10.4.2 Interior of the reactor

Both containers in Figure 10.4.4 are equipped with a cooling coil(4), thermocouple insertion tube(5), and an electric heater(6). Each of these items is arranged symmetrically about the center of the container. Both containers are also equipped with a pressure gauge(7). In addition, the evaporator/condenser has a water level gauge(8) and the reactor has an auxiliary external heater(9). The auxiliary heater consists of nichrome wire wound around the reactor's outer surface. The two containers are connected to each other by stainless steel piping via valve V2. A vacuum pump(3) is used to obtain the proper pressure level in the reactor. The equipment is insulated with an adiabatic material to reduce heat loss. Figure 10.4.5 shows details of the reactor's construction.

For the reaction system employed, the temperature $T_M$ shown in Figure 10.4.2 is in principle about 640K. At this temperature, the pressure of the saturated steam in the evaporator is 27.5MPa. The present experimental apparatus was not designed for such high pressures. For this reason the evaporator temperature was kept below about 430K for this experiment. The CaO specimen employed in the experiment was prepared by calcining the limestone from Hiroshima (the purity of $CaCO_3$ is 99% and the particle diameter 0.7~1.0mm) at about 1223K in the electric furnace [10.4.4]. The CaO layer in the reactor was 110mm thick and rested on a 30mm thick asbestos insulating sheet.

For the heat release experiment, the temperature of the specimen layer and the steam pressure within the evaporator were adjusted to their proper values using heaters(6) and (9) shown in Figure 10.4.4. Valve V2 was then opened to allow pressurized steam from the evaporator to enter into the reactor causing an exothermic hydration. During the reaction period, the temperature changes in time which took place within the specimen layer was measured with a thermocouple and recorded automatically. Changes in the evaporator/condenser water level were also measured.

In the experiment on heat storing, cold water was passed through the condenser/evaporator cooling coil(4) to maintain a temperature of $T_L$. At the same time, the reactor was heated by heaters(6) and (9) to cause $Ca(OH)_2$ to undergo an endothermic dehydration. The steam produced by this reaction was condensed in the condenser/evaporator and as before the temperature change in time within the specimen layer and the change in water level within the condenser were measured and recorded.

### 10.4.5 Results of Chemical Heat Pump Experiments

(i) Heat release mode
Figure 10.4.6 shows the temperature changes in time at various radial positions of the CaO packed bed (the axial position is 1/2 of the bed height) in the reactor during the exothermic hydration of CaO with steam. For the initial condition of this experiment, the evaporator was kept at the same temperature of 444K (the steam pressure is 0.142MPa) as the reactor. As seen in Figure 10.4.6, as soon as pressurized steam is introduced into the reactor, the temperature rises instantaneously and maintains a fairly constant temperature of about 800K (pseudo-equilibrium temperature) for 120 minutes, after which it slowly decreases to its initial temperature. In this case, the reactor wall is substantially the heat discharging surface, so that after the

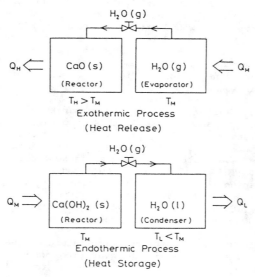

Figure 10.4.3 Operation modes of chemical heat pump

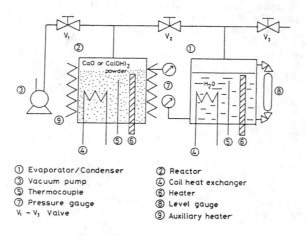

① Evaporator/Condenser  ② Reactor
③ Vacuum pump  ④ Coil heat exchanger
⑤ Thermocouple  ⑥ Heater
⑦ Pressure gauge  ⑧ Level gauge
$V_1 - V_3$ Valve  ⑨ Auxiliary heater

Figure 10.4.4 Schematic diagram of the experimental unit

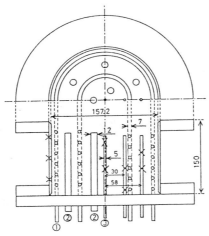

① Coil heat exchanger
② Rod heater
③ Thermocouple lead
X Thermocouple

Figure 10.4.5 Details of the reactor

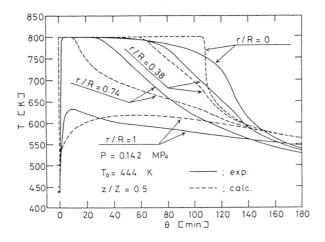

Figure 10.4.6 Temperature changes in time during heat-release step (hydration of CaO)

temperature of each part of the specimen layer rises uniformly and instantaneously up to the pseudo-equilibrium temperature, it eventually drops from the point closest to the wall to the center of the reactor. From a relatively large temperature gradient in the radial direction, it is supposed that the heat transfer capacity of the specimen bed is not so sufficient. In order to make a solution to this problem, we attempt to enhance the heat transfer rate through the specimen bed by applying a passive-type heat transfer augmentation technique [10.4.5, 10.4.6].

As the result of the experiments carried out by changing the steam pressure (evaporation temperature), it was verified that the pseudo-equilibrium temperature of the hydration of CaO shifted to a higher temperature with the increase of steam pressure in the evaporation temperature range between 378K and 428K. As shown in Figure 10.4.7, the obtained reaction temperature were well correlated with the steam pressure by the ln P - 1/T line measured by Halstead and Moore [10.4.7].

(ii) Heat storage mode
Figure 10.4.8 shows the temperature changes occuring in the reactor at various radial positions during the endothermic dehydration of $Ca(OH)_2$. The initial temperature of the reactor $T_O$ and the condenser water temperature $T_C$ in this case were 875K and 277.5K, respectively. At the first step of the research, electric heaters were used in this experiment for the substitution of high-temperature heat sources for the regeneration of CaO. In this point, it must be said that the present heat storage experiment does not strictly correspond to the chemical heat pump experiment, since the chemical heat pump ought to work with heat sources not with electrical power. The present study, however, is located to ascertain the possibility whether CaO can be regenerated in the heat pump unit continuously.

As seen in Figure 10.4.8, no sooner had valve V2 been opened than the temperature of the $Ca(OH)_2$ bed drops instantaneously from the initial temperature of 873K to about 693K, owing to a relatively high reaction rate and a large quantity of reaction heat of the dehydration of $Ca(OH)_2$. The $Ca(OH)_2$ bed is kept at a fairly constant temperature during the reaction by absorbing the reaction heat from electrical heaters continuously. This pseudo-equilibrium reaction temperature is nearly the same as that appeared in the hydration of CaO, but there exists a slight difference between the two. This may be caused by a local temperature rise in the condenser; the cooling capacity of the condenser is not enough to cope with such an abrupt phenomenon in condensing the superheated steam released from the reactor. These differences in pseudo-equilibrium temperature of the hydration of CaO and that of the dehydration of $Ca(OH)_2$ are shown in Figure 10.4.7. From this point of view, the development of a high performance evaporator/condenser is considered as one of the most important keys to realize the chemical heat pump.

Conclusion

A chemical heat pump which works in a relatively high temperature region over 773K by using the combination of the hydration/dehydration of $CaO/Ca(OH)_2$ and the evaporation/condensation of water was proposed,

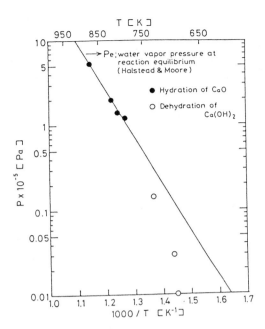

Figure 10.4.7 Relation between the equilibrium-pressure and temperature

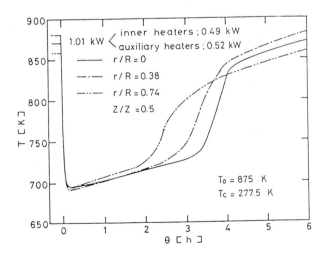

Figure 10.4.8 Temperature changes in time during heat-storage step (dehydration of $Ca(OH)_2$)

and the experiments with a manufactured laboratory scale chemical heat pump unit were conducted.

As a result, under the present experimental condition, we could produce a high temperature level heat of which the temperature is above 800K with the steam of about 428K. The proposed chemical heat pump possesses a limited extent in its application, because the regeneration temperature of CaO is a relatively high. However, this chemical heat pump is considered applicable to the utilization of high temperature industrial waste heat, the cogeneration system, atomic power plants, the thermal energy storage system and so on.

The authors are now in the second step in developing this chemical heat pump with a Mark II experimental unit. The obtained results will be presented on some other occasion.

References

[10.4.1] Society of Chemical Engineers, Kanto Branch, Japan, New Development of Thermal Energy Utilization, p.97, 1983.
[10.4.2] Kanzawa, A. and Y. Arai, Thermal Energy Storage by the Chemical Reaction - Augmentation of Heat Transfer and Thermal Decomposition in the CaO/Ca(OH)$_2$ Powder, Solar Energy, vol.27, no.4, pp.289-294, 1981.
[10.4.3] Matsuda, H., M. Miyazaki, M. Hasatani and M. Yanadori, Temperature Response of Solid Particle-bed in High-Temperature Chemical Heat-Pump Using Ca(OH)$_2$/CaO Reversible Thermochemical Reaction, Kagaku Kougaku Ronbunsyu, vol.14, no.6, pp.769-778, 1988.
[10.4.4] Matsuda, H., T. Ishizu, S. K. Lee and M. Hasatani, Kinetic Study of Ca(OH)$_2$/CaO Reversible Thermochemical Reaction for Thermal Energy Storage by Means of Chemical Reacton, Kagaku Kougaku Ronbunsyu, vol.11, no.5, pp.542-548, 1985.
[10.4.5] Matsuda, H., M. Miyazaki. M. Hasatani, M. Yanadori and Y. Ezaki, A Basic Study on High-Temperature Chemical Heat-Pump by Use of Ca(OH)$_2$/CaO Reaction, Preprint of the 25th Japan Heat Transfer Symposium, pp.148-150, 1988.
[10.4.6] Ogura, H., H. Matsuda, M. Hasatani, M. Yanadori, M. Hiramatsu and T. Inoue, Studies on Heat Transfer Characteristics of Solid Reactant Bed in Chemical Heat Pump by Use of Ca(OH)$_2$/CaO Reaction System, Preprint of the 22th Autumn Meeting of Chemical Engineers, Japan, SM213, p.630, 1989.
[10.4.7] Halstead, P. E. and A. Moore, The Thermal Dissociation of Calcium Hydroxide, J. Chem. Soc., pp.3873-3875, 1957.

# 11. EFFICIENCY IMPROVEMENT OF ENERGY-RELATED FACILITIES FROM AN ENVIRONMENTAL VIEWPOINT

**Kunio Hijikata**
Tokyo Institute of Technology
Department of Mechanical Engineering Science
2-12-1, Ohokayama, Meguro-ku, Tokyo 152 Japan

## 11.1 Introduction

It is not incorrect to state that all major environmental problems, such as the greenhouse effect, destruction of the ozone layer from CFC's, acid rain due to air pollution by NOx and SOx, etc., are caused by excessive industrial and residential energy consumption. Considering the finite world energy resources and limited global space, the day might be already upon us in which the total amount of energy consumption in the world should be reduced. To maintain a high living standard without increasing energy consumption, waste energy recovery and energy conservation are vitally important. Since the technological aspects of energy conservation are stated in the previous sections, current efforts concerning effective energy utilization in the residential and commercial arenas are described from the standpoint of available energy. Future directions in energy research, considering environmental aspects and energy resource conservation, are also discussed.

## 11.2 Conservable energy and consumable energy

One of the most fundamental laws in physics is the first law of thermodynamics, which states that the total amount of energy remains constant during any energy conversion and/or transmission. For example, examine the global energy balance shown in Fig. 11.1, which is the most important consideration for mitigating the greenhouse effect [11.1]. Twenty-five % of the incoming solar energy is absorbed in the atmosphere, and the same fraction of energy is reflected back to space. The remaining 50 % travels through the atmosphere and reaches the earth. Ten % of it, which corresponds to five % of the total incoming solar energy, is reflected at the surface of the earth and the remaining 90 % is absorbed. The earth also absorbs/emits radiative energy from/to the surrounding atmosphere. These transferred energies correspond to 88 % and 104 % of the incoming solar energy, respectively. Furthermore, the earth's thermal energy is lost to the atmosphere by convection, and both the earth and the atmosphere emit energy to space through radiation. The radiation emitted by the atmosphere and absorbed by the earth is the cause of the *greenhouse effect*.

At any horizontal plane in Fig. 11.1, the inlet energy (the energy flow to the earth) balances with the outlet energy. Outside of the atmosphere, the energy flow to the earth consists only of the incoming solar radiation and is equal to the energy leaving the earth; that is, the sum of the emitted atmospheric radiation and the solar radiation reflected from the atmosphere and from the earth. Even at the surface of the earth, the incoming energy is exactly balanced by the outgoing energy.

The increase of $CO_2$ content in the air increases the thermal radiation from the atmosphere to the earth, bringing about a rise in the earth's temperature. At the same time, however, the radiation from the earth to the atmosphere also increases to compensate for the greater amount of incoming energy. This suggests that if the heat transfer rate from the earth to the atmosphere, and then out to space, could be increased, the greenhouse effect would be cancelled without having to reduce the $CO_2$ content in the air.

Currently, our total energy consumption equals one third of the total biomass energy produced by the incident solar energy, as shown in Fig. 11.2. This means that we must expend non-renewable energy resources, like fossil fuels, to maintain our energy consumption rate. This practice is using up our global energy "savings". Thus, considering our present condition, we should make efforts to reduce our energy consumption, regardless of the effects of increasing $CO_2$ content on the environment.

Figure 11.1 shows that energy is always conserved and is never lost, which is inconsistent with the idea of energy saving or effective energy use. Usually the word 'energy' has two meanings: *conservable* energy and *consumable* energy. Exactly speaking, the total energy, like that considered in an energy balance, refers to the conservable energy, but the consumable energy, also called the *available* energy, is the energy that can actually be used to do useful work. For example, the surrounding atmosphere has *conservable* energy but no *available* energy.

The available energy A refers to the maximum possible work output of a system. For an open energy system, it is given by [11.2]

$$A = H - H_o - T_o(S - S_o) \tag{11.1}$$

where H and S are the enthalpy and the entropy, respectively, and the subscript o means the surrounding environmental condition. In other words, the available energy A is the difference between the conservable energy, $H - H_o$, and the energy spent to maintain the disorder of the system, $T_o(S - S_o)$, where the entropy S is essentially a measure of the disorder of a system. From Eq. (11.1), it is clear that the system available energy is zero at the environmental condition. If there is just a small temperature difference between the system and its surroundings, A is proportional to the square of the temperature difference, indicating that as the temperature decreases, A rapidly goes to zero.

The conservable energy Q is simply

$$Q = H - H_o \tag{11.2}$$

Q also becomes zero at the surrounding condition, but in such a situation that energy is not lost; rather, it increases the energy of the surrounding medium. Since the surrounding medium has infinite heat capacity, its temperature does not change, but the total energy of the surrounding medium is still increased. The available energy, however, is reduced due to the energy conversion and transmission. Plainly speaking, this means that the energy offered to us is always more than we can accept, or that we can use. Therefore, we can say that our lives are spent consuming the available energy.

For example, if a cup of water whose temperature is $T_o + DT$ is mixed with the same amount of water having the surrounding temperature $T_o$, the final temperature of the mixed water is $T_o + DT/2$. This temperature difference between the water and the surrounding environment is only half of the initial temperature difference, but the total amount of water is doubled and the total thermal energy therefore does not change. However, the total available energy obtained from Eq. (11.1) is halved, even though the total amount of water is doubled. This example clearly shows that the available energy is consumed by mixing, transfer, exchange, and conversion processes. These processes are all irreversible, and our daily life can be considered a succession of irreversible processes.

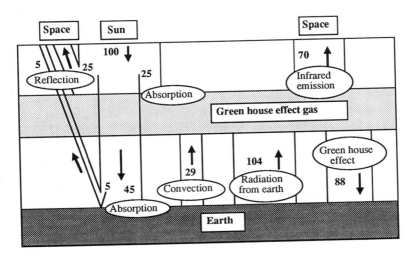

Fig.11.1 Energy Balance on the Earth

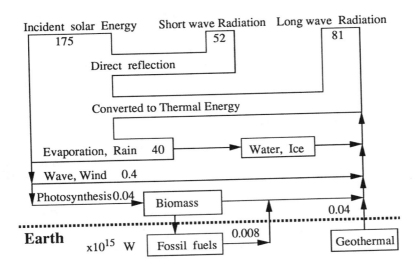

Fig.11.2 Photodsynthesis Energy and Fossil fuel Consumption

From these arguments, it is very important to consider the ratio of the available energy to the thermal energy, or in other words how much of the total energy can be converted into work. For simplicity, we assume $T > T_0$ and $P = P_0$, where P is pressure. Under these conditions, the ratio of the available energy to the total conservable energy, which is hereafter referred to as the *available energy rate* e, is given as

$$e = 1 - T_0(S - S_0)/(H - H_0) \qquad (11.3)$$

The available energy rate for air, as a function of temperature, at 101.3 kPa pressure is shown as the solid curve in Fig. 11.3. The available energy rate e is zero at the surrounding temperature and increases with temperature. The figure indicates that even for air at solar temperatures (5770 K), e is less than one, suggesting that the available energy rate must always be less than one, i.e., the *quality* of thermal energy is relatively low compared to, for example, that of electrical energy. Here, energy quality refers to both its available energy and to its ease of use.

In the figure, the working temperatures of typical steam and gas turbines for power stations are also shown for reference. The available energy rate corresponds to the maximum efficiency of energy conversion to useful work, but it is not equal to the Carnot cycle efficiency, which is generally considered to be the maximum possible efficiency. In the Carnot cycle, the temperature of the high temperature heat source is kept constant, but when considering the available energy rate, the system temperature decreases due to the energy flow out of the system. Since the available energy rate decreases with decreasing temperature, in all processes involving a temperature change, we can obtain useful work corresponding to the available energy change. The process in which the temperature decreases, but results in no output work, is the most ineffective process.

## 11.3 Flow of Thermal Energy

The energy resources in Japan, and the recent trends of energy consumption in Japan, are shown in Fig. 11.4 [11.3]. Most of the energy resources are first converted to thermal energy, then changed into mechanical and electrical energy, and then finally utilized in the final stages. Numbers given in the map of Fig.11.4 indicate the ratio of that energy to the total energy consumption. In the final stage, an item of loss is given, which is the difference between the total used available energy and the total output work in the energy conversion process. This is energy consisting of conservable energy, but no available energy. Thus, in all conversion processes some loss is generated. Strictly speaking, all other items in the final stages are also considered losses after finishing the objective energy use.

Efficiencies of typical energy conversion processes are shown in Fig. 11.5 [11.4], although they do not strictly correspond to the available energy rate because the definition of the efficiency is somewhat arbitrary. It is clear that the lowest efficiency processes are mainly those concerned with conversion from thermal energy and to/from light energy. Since most of the energy is in the form of thermal energy, as shown in Fig. 11.4, improving the thermal energy conversion process is of primary concern. The main reason for the low efficiency of thermal energy conversion can be explained by considering the available energy rate.

As shown in Fig. 11.3, the available energy rate of electric power is 100%, because all of the electrical energy can be converted to work in an ideal situation. Electrical energy is thus the highest quality energy available today. Fossil fuels, like natural gas, petroleum and coal, have about 95% available energy rates. However, the effective available energy rates of fossil fuels are much lower, because to make thermal energy from these energy resources, the maximum temperature must be kept below 1500 °C, from engineering restrictions of the combustor wall materials and production of pollutants like NOx and SOx. The available energy rate of air at 1500 °C is only 65%, and therefore 30% or more

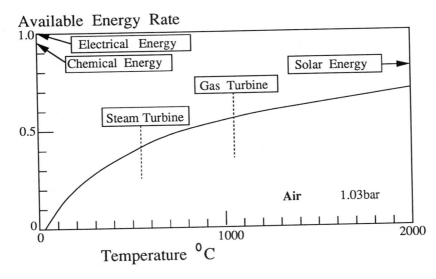

Fig. 11.3 Available Energy Rate againt Temperature

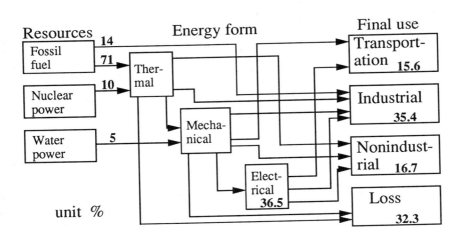

Fig.11.4 Energy flow with conversions

available energy is lost during the combustion process. Thus, as long as a combustion gas is used as a high temperature heat source, the conversion efficiency will never exceed 65%, and this is why the conversion efficiency using thermal energy is so small.

It is usually said that effective use of high temperature energy sources is most important for improving the total energy efficiency. However, Fig. 11.3 shows that this is not quite true, since the gradient of the available energy rate with respect to temperature is very small in the high temperature region. For example, for air the available energy rate increases 0.15%/°C below 100 °C, but it becomes about one tenth of that value at 1500 °C.

A reduction in the temperature drop of 1 °C in the heat transfer process is more effective in the low temperature range compared with the high temperature region, and this improvement is easier from a technological viewpoint. Therefore, efforts towards improving low temperature thermal energy utilization, which is mainly used in nonindustrial fields, help to solve our energy/environmental problems.

## 11.4 Available Energy Loss by Mixing

The available energy rates of fossil fuels are about 95%, which means that fossil fuels have the capability of generating very high temperatures. In a real combustion process, however, the maximum temperature never exceeds 2000 °C, and in some cases it is below 1000 °C, especially for a lean mixture of fuel and air. Although the available energy rate of 1000 °C air is about 55%, this reduction of the available energy rate comes about because of the decrease in temperature, and not from the mixing of lean fuel and air. The mixing process does contribute to an additional available energy loss, along with heating of already burned products.

For a qualitative discussion, the reduction of available energy rate by mixing is shown in Table11.1. A 0.1% diluted methane in air mixture still has 97.9% of the available energy of pure methane (for the same amount of methane). Thus, the diluted fuel is still capable of generating high temperatures, but only by employing a special combustor. Since these lean fuels have very small reaction energies, combustion must occur at high temperature. Therefore, the fuel and air mixture must be heated before combustion, as shown in Fig. 11.6 by a heat exchanger. Only the generated heat is removed from the combustor, and the remaining heat, which is introduced by the heat exchange process, is used as the thermal energy source of the heat exchanger. This system, as described, can operate either with a perfect heat exchanger having 100% efficiency, or where much more heat is generated than is exchanged. The similar system shown in Fig. 11.6 is effective for improving the energy efficiency. Echigo [11.5] proposed the same system but without the heat exchanger, where the thermal energy from the combustion gas is transferred to the unburned mixture by thermal radiation. A more practical example of this system is in air preheating of combustion gases.

In a huge energy system like an electric power generation plant, an available energy analysis was carried out [11.6] and a similar type of heat exchanger was adopted. However, very few small energy systems for nonindustrial use have adopted this heat exchange system. Only in one special case of a small energy system has the system shown in Fig. 11.6 been adopted for energy conservation. That example is an animal. For small animals, it is very difficult to get enough food to maintain their body temperature. The heat loss from their protuberances, like tails, legs or fins, are very large, especially if they live in a cold climate. Therefore, they have special heat exchangers called *wonder nets* [11.7]. In these systems, the combustor, the heat exchanger, and the working fluid correspond to the heart, the wonder net, and blood. High-temperature arterial blood is cooled by the wonder net and transmitted to the protuberant parts. The cooled venal blood is heated by the wonder net and returned to the heart. By this system, animals can reduce their consumption of energy, which is very important since they obtain all their energy

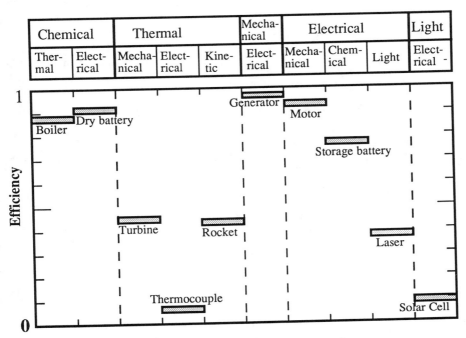

Fig. 11.5 Efficiencies of various energy conversions

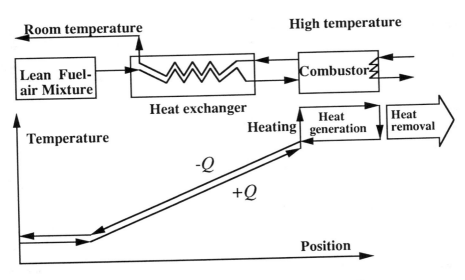

Fig.11.6 Effective use of energy by using a heat exchanger

from their food. This example shows us that improvements in energy efficiency are still possible even in small energy systems, like those in the nonindustrial energy field.

## 11.5 Nonindustrial Energy Use in Japan

To discuss the possibilities of improving energy efficiency in the Japanese nonindustrial field, the present energy consumption situation must first be examined. Generally, in industrialized countries the percentage of commercial and residential energy utilization in the total energy consumption follows this pattern: it initially decreases with an increase in the national income per capita, reaches a minimum, and finally increases as the national income increases. The initial reduction in the percentage of nonindustrial energy consumption is caused by industrialization, which means that the percentage of industrial energy consumption rises. However, once the national income becomes relatively high, people expend large amounts of energy to increase their standard of living, and thus the energy consumption rate in the residential field becomes larger. In the industrialized countries that maintain a high standard of living, the industrial field has changed from producing low-value goods requiring large energy consumption, to high-value goods requiring relatively low energy consumption. Canada, the United States, West Germany, and Sweden belong to this category of nations [11.8]. As for Japan, residential and commercial energy consumption are now reaching a minimum. Therefore, the nonindustrial energy consumption in Japan will increase more and more in the near future, and thus energy saving in this area is becoming the most important subject of energy conservation research.

This trend can be seen in Fig. 11.7, where the change in nonindustrial energy consumption versus time in years, and the percentage of each energy resource, are shown [11.9]. The consumption of kerosene is nearly constant, but natural gas and especially electric power consumption are increasing. This trend indicates that people want convenient and easy-to-use energy. The figure also shows energy end uses. The energy consumption for heating and for hot water supply are predominant, which suggests that future efficiency improvements should be expected. Electrical energy consumption for lighting and for electrical equipment are both increasing. The energy saving in this field is stated below.

The temperature range of used thermal energy is shown in Table 11.2 for the industrial, commercial, and residential fields [11.8]. The amounts of energy consumed by commercial and residential users are nearly equivalent. For residential consumption, the used thermal energy is generally below 100 °C; for commercial use, it is between 100 and 300 °C; and for industrial use, the temperature range is over 300 °C. It is clear that efficiency improvement in the low temperature region is most important for commercial and residential energy users.

## 11.6 Present Electrical Energy Use

Since electrical energy has the highest quality and the highest available energy rate, the biggest effort has been made to improve the energy efficiency of electrical facilities. In Fig. 11.8, the consumed electrical power of a refrigerator is plotted against the production year. The major areas of improvement are also shown in the figure [11.8]. After 1985, the consumed electrical energy for a given size of refrigerator became constant and thus no more drastic improvements can be expected. The same tendency is observed in various other electrical equipment, like Color TV's, air conditioners, etc. [11.9].

Another important consumer of electrical energy is lighting. The power used for lighting is increasing more and more with an increasing national income. Usually, as shown in Fig. 11.3, the efficiency of energy conversion from/to light is very low, but great efforts have been spent on improving lighting efficiency. For an incandescent lamp, the invention of the halogen lamp improved the efficiency [11.10], as shown in Fig. 11.9, but still the efficiency is only one fifth of the theoretical value. The efficiency can also be enhanced

| $CH_4$ | Available Energy | |
|---|---|---|
| 100.0 % | 829 kJ/mol | 0.0 % |
| 1.0 | 817 | 1.4 |
| 0.1 | 811 | 2.1 |

| $CO_2$ | Available Energy | Increasing Rate |
|---|---|---|
| 0.03 % | 0 kJ/mol | 0.0 % |
| 100. | 20 | 5.0 |

Table 11.1 Available Energy Change by Mixing

| Industrial use | 0-100 | 100-200 | 200-300 | 300- C |
|---|---|---|---|---|
| Paper manufacturing | 0 | 10 | 5 | 85 |
| Chemical industry | 3 | 8 | 10 | 79 |
| Ceramic industry | 1 | 8 | 4 | 87 |
| Steel making | 0 | 1 | 1 | 98 |
| Other industries | 3 | 36 | 6 | 55 |

| Commercial use | 0-100 | 100-200 | 200-300 C |
|---|---|---|---|
| Electric heating | 20 | 40 | 40 |
| Coal | 20 | 40 | 40 |
| Gas | 25 | 40 | 35 |
| Others. | 20 | 40 | 40 |

| Residential use | 0-100 | 100-200 | 200-300 C |
|---|---|---|---|
| Heating | 100 | 0 | 0 |
| Hot water supply | 100 | 0 | 0 |
| Cooking | 30 | 30 | 40 |

Table 11.2 Temperature range of energy using in industrial, commercial and residential fields

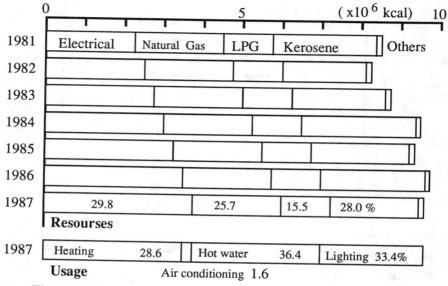

Fig. 11.7 Energy consumption for residential use in Japan

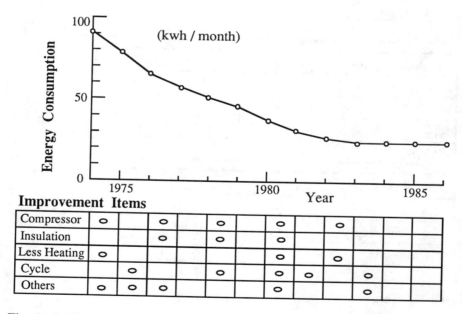

Fig. 11.8 Change of power consumption of a refrigerator for residential use

by coating the lamp glass with an optical filter, which transmits only visible light, and thus does not permit infrared light, which has thermal energy, to go out through the bulb. A combination of this halogen lamp with this special reflector (named a 'dichroic' mirror), which, as shown in Fig.11.10, reflects visible light and allows all other energy to go through, increases the total efficiency. Additionally, the discovery of the three-color fluorescent light has also improved the efficiency, but the efficiency is still low and a new concept in lighting is necessary for further significant improvement. Since new concepts occur infrequently, this suggests that the total energy consumption for lighting will increase in the future.

## 11.7 Effective Use of Energy in the Commercial Field

As shown in Fig. 11.3, since a reduction in available energy rate is caused by energy transfer or conversion, a suitable combination of energy conversion facilities will increase the system efficiency. An especially important example is when the system load changes on a day-to-day basis and also on a year-to-year basis. In such a case, the combination of two or more smaller facilities are effective. For commercial users, the system load is often big enough to be divided among two or three facilities. Figure 11.11 shows a large refrigeration system consisting of seven different kinds of refrigerators [11.11]. By changing the arrangement of these refrigerators, a high COP (coefficient of performance) can be obtained over a wide range of thermal loads, as shown in Fig. 11.12. In the figure, eleven different operation modes are realized by changing the combination of seven refrigerators.

Even in a small refrigeration system, the daily output power can be varied by using a inverter system. An inverter system is used to change the frequency of the input electricity to optimize performance. Figure 11.13 gives a comparison between the consumed power of an ordinary refrigerator with that driven by an inverter system [11.12]. By using an inverter system, an electrical power savings of 20% can be achieved. The system operates constantly at the optimum condition for energy saving. These improvements have only recently been realized by the development of cheaper power control circuits. This inverter control system will be applied to many different applications for energy conservation purposes.

## 11.8 Effective Use of Energy in the Residential Field

Energy consumption in the residential area is about half of the total nonindustrial use, which is about 16.7% of the total energy consumption in Japan. This is not negligible from an energy conservation viewpoint. Natural gas, LPG, and kerosene are used for cooking, hot water supply, and heating, where the operating temperatures are relatively low. The available energy rates are roughly 10 to 20%. If electrical energy is converted to this low-temperature thermal energy, 80 to 90% of the total required energy can be obtained from the surrounding atmosphere by utilizing a heat pump, leaving only the remaining 10 to 20% to be supplied as electrical energy. Thus, there still exists a great possibility of energy saving in the residential field.

Another energy-saving idea is the so-called co-generation system of electrical power and heat. By using this system, a reduction in air pollution is also expected, because while reduction of air pollution is impossible in small heating furnaces in individual houses, it is possible in large systems. Most of the heat supply systems in northern European countries were constructed for this purpose. In Japan, these systems were mainly constructed for the effective use of waste thermal energy. Both objectives are acceptable from the standpoint of energy conservation and environmental concerns.

Energy resources for the public heat supply are different in different countries [11.13]. Over 25% of the residential thermal energy is supplied by district heating in northern countries like Sweden, Finland, and Denmark. Even in the United States, France, and Austria, 2 to 5% of the thermal energy is supplied via pipe line. In Japan, however, the

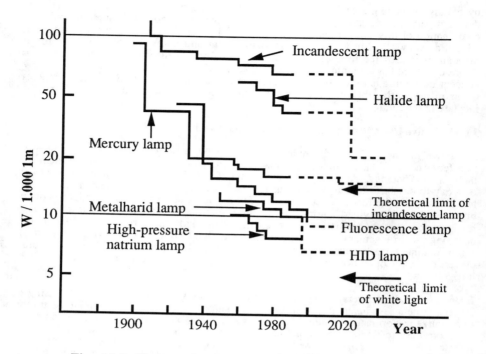

Fig. 11.9 Change of consumption of lighting power

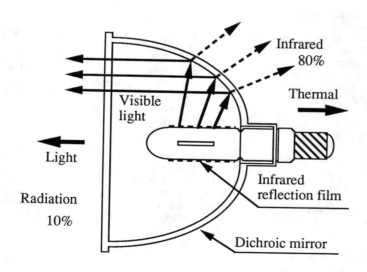

Fig. 11.10 New type lamp for energy saving

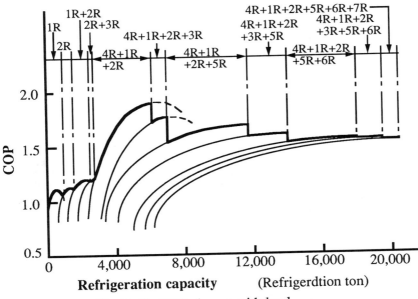

Fig.11.11 COP change with load

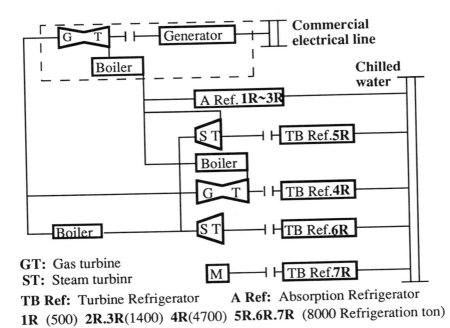

**GT:** Gas turbine
**ST:** Steam turbinr
**TB Ref:** Turbine Refrigerator      **A Ref:** Absorption Refrigerator
**1R** (500) **2R.3R**(1400) **4R**(4700) **5R.6R.7R** (8000 Refrigeration ton)

Fig. 11.12 A refrigeration system for energy saving

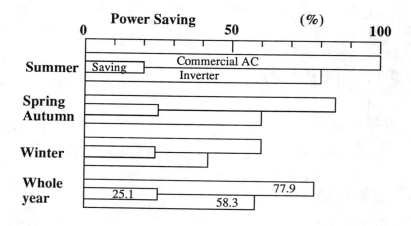

Fig. 11.13 Energy saving of a refrigerator by inverter

percentage of residential heating supported by a district heating system is negligible. The relatively low demand of thermal energy in Japan is one reason why the heat supply system did not develop, but another reason is that the administrative agencies have never promoted these systems. For example, in Japan, a district heating system does not have priority to use the ground directly beneath a road.

Except for a district heating system, for an individual house, a heat pump using a fossil fuel as an energy source is the ideal system, where high-temperature drain water, river water, or atmospheric air can be used as the heat source. Since fossil fuels have a high available energy rate, the thermal efficiency will increase drastically if a heat pump cycle operated directly on fossil fuels is developed. An absorption heat pump, a heat pump combined with a gas engine, and a combined system of two Stirling cycles named the Vuilleumier cycle [11.14] are newly proposed systems, but they still require further improvement.

## 11.9 Conclusions

Energy conservation from a technological perspective has already been discussed in the above sections. I'd like to conclude by briefly discussing energy conservation from a political perspective. Energy saving can be achieved by changing the national industrial structure from one that produces high-energy-consuming goods to one that produces low-energy-consuming goods, and by an effort to make cheaper, smaller, but higher quality electrical equipment. For this reason, all of the ideas for energy saving can be realized only in the developed countries. Therefore, from the viewpoint of global environmental problems, technology transfer and the transfer of technical assistance for energy management from developed to undeveloped countries are vitally important.

## References

[11.1] S.H.Schneider, "Climate modeling", *Scientific American*, Vol.256, No.5, (1987) pp.72-81

[11.2] Japanese Industrial Standard "General Rules for Energy Evaluation Method by Available Energy", Z9204, 1980

[11.3] E. Yamada "World's Energy Resources and Japan's Energy Utilization", *Energy and Resources*, Vol.9, No.4, (1988) pp.19-26

[11.4] C.M.Summers "The Conversion of Energy", *Scientific American*, Vol.225, No.3, (1971) pp.148-163

[11.5] R. Echigo et al.., "Analytical and Experimental Studies on Radiative Propagation in Porous Media with Internal Heat Generation", *8th International Heat Transfer Conference,* Vol.2 (1986) pp.826-832

[11.6] K. Nishikawa, "Available Energy Account of Steam Power Plant", *Production Research Institute Report in Kyushu University*, 67, (1987) pp.7-21

[11.7] H. Ohara, "Why does a crane sleep standing on one leg?", *Soshisha*, (1987)

[11.8] Y. Muroda "Characteristics of nonindustrial energy consumption in Japan", *Energy and Resources*, Vol.8, No.4 (1987) pp.19-28

[11.9] Energy Conservation Center, "Energy Conservation Handbook", (1989)

[11.10] T. Hanada and Y. Yuge, "History of Lighting Technology", *Energy Review*, November (1988) pp.20-25

[11.11] M. Sakauchi, "District Air Conditioning i n Urban Renewal ", *Hitachi Review*, Vol.69, No.85, (1987)

[11.12] S. Suzuki, "Inverter Refrigerator for Outdoor Use", *Toshiba Review*, Vol. 42, No.35 (1987)

[11.13] T. Hashimoto, "District Heating in Europe and U.S.", *Energy and Resources*, Vol.10, No.1 (1989) pp.18-26

[11.14] S. Fujimaki et al.., "Development of Now Air Conditioning System using Vuilleumier Cycle", *Proceedings of Japan Refrigerant Society Meeting*, (1989) pp.65-68

# INDEX

Acid rain, 175–196
   (*See also* Nitrous Oxide; Sulphur oxide)
Air-conditioners, 11
   (*See also* Chloroflourocarbons)
Aircraft, 99, 155
Alternative energy, 197–233
   (*See also* specific types)
Ammonia, 166
Atmosphere:
   as closed system, 22, 23, 25
   paleo-atmosphere, 24, 26–27
   structure of, 145, 149
   (*See also* specific gases)
Automobiles:
   carbon dioxide and, 41–42
   diesel engines, 4
   emission controls for, 41–42
   hydrogen-fueled, 101
   metal hydrides, 109–121
   methanol fuel, 3, 62
   nitrous oxide, 61, 188, 192
   UCLA SEED Rally, 123–126

Bioact EC-7, 165
Biological systems:
   biomass burning, 27–30
   $CO_2$ fertilizer effect, 26
   hydrogen production, 88
   photosynthesis, 40–41, 88
   small system energy conservation, 330
   (*See also* Environmental protection;
      Methane; Oceans)

Calcium, and $CO_2$, 43–46
Carbon, 8–9, 63–68, 235
Carbon dioxide, 3, 5, 16, 21, 206
   absorption of, 3, 40, 44
   annual emission, 2, 26, 64
   atmospheric levels, 24, 26, 40, 43
   automobiles and, 41–42
   cogeneration, 311
   cost of separation, 59–61
   cycle of, 25, 43
   economic growth and, 2
   electrical power and, 15
   emissions by country, 14–15
   forests and, 21, 26, 40–41
   fuel sources and, 12–13
   global energy use and, 3
   global warming and, 24–26
   hydrocarbon-fired plants, 40–61
   in Japan, 15, 20
   levels for, 24, 26, 40, 43
   liquified, 41–42, 48–49, 51–54
   in ocean, 43–47
   photosynthesis and, 40–41
   radiation-controlled combustion, 64–81
   separation of, 40–41, 55–61
   solidified $CO_2$, 51, 64–81
   sources of, 43
   super heat pump, 284, 285, 301–303
   (*See also* Chloroflourocarbons; Methane)
Catalytic processes, 172–173, 266–267
Ceramic Gas Turbine (CGT) project, 265,
   267–277, 308
Chapman mechanism, 147
Charcoal production, 8–9
Chemical heat pump, 313–324
   cogeneration with, 313
   load-leveling, 313
   Mark II experimental unit, 324
   reaction candidates, 314
China, geothermal energy, 216
Chloroflourocarbons, 3, 31–33, 145–174
   alternatives to, 155–157, 160–166
   atmospheric levels, 147–149
   catalytic processes for, 172–173
   demand for, 155–156
   destruction of, 167–174
   geothermal energy, 216
   global warming effect, 152–153
   in Japan, 168
   kinds of, 145
   lifetime of, 147–149
   propellant policies, 155
   recovery of, 156–157, 167
   regulations for, 153–156
   supercritical water, 171–172
   super heat pump, 299
   U.S. levels, 168
   (*See also* Ozone)
City gas systems, 101
Cleaning agents, 164–165
Climate change (*see* Global warming)
Club of Rome studies, 3

Coal:
    gasification/liquification, 51, 198, 256
    industrialization and, 8, 9
    mining of, 27–29
    (*See also* Fossil fuels)
Cogeneration, 100, 305–312
    carbon dioxide from, 311
    CGT project, 308
    chain cycle, 308
    chemical heat pump, 313
    defined, 305
    demand for, 308
    electricity usage and, 306, 310
    environmental aspects of, 310–312
    fuel cells, 308
    in Japan, 306, 310
    Moonlight Project, 308
    nitrous oxide, 311
    regulation standards, 310–311
    Stirling engines, 311

Deserts, 206
Diesel engines, 4
DHP program, 217

Eastern Europe, energy demands, 10–11
Economics:
    consumer shifts, 332
    energy demands and, 17, 256, 332
    environmental trade-offs, 6
    low energy products, 332
    market allocation model, 198
    oil prices and, 3
Efficiency, of energy use, 323–339
    animals and, 330
    electricity and, 332
    fossil fuels and, 330
    fuel/air mixing and, 330
    low temperature thermal energy and, 330
    power transmission and, 18
    preheating and, 330
    public heating systems and, 335
    refrigeration inverter system and, 335
Electric power, 3, 256–281, 310
    carbon dioxide and, 261–262
    catalytic combustors and, 266–267
    ceramics for, 265, 268–277
    coal gasification, 51, 198, 256
    cogeneration, in Japan, 12, 306
    demand growth of, 256
    fluidized bed combustion, 256
    fuel cells and, 277–280
    gas turbine technology, 257
    history of, 256–262
    load control, 258–260
    methanol and, 256–257
    nitrous oxide, 261–262, 266–267
    production temperatures, 256
    sulfur oxide, 261–262
    superalloys, 265
    thermal efficiency, 257–262, 265
    turbine systems, 262, 264–265
Electrolysis, hydrogen from, 62, 85–86, 250–252
Electron irradiation emissions, 193–195
Energy conservation, 3, 5–6, 256–324
    (*See also* Electric power; Super Heat Pump)
Enteric fermentation, 27–29
Environmental protection:
    alternative energy and, 197–199
    cogeneration and, 310–312
    energy conservation, 299–303
    heat pump and, 284–285
    UNEP program, 1, 40, 155
    (*See also* specific topics)

Foaming agents, CFCs and, 163–164
Forests, 2
    acid rain and, 195
    carbon dioxide, 21, 26, 40–41
    nitrous oxide, 29–30
Fossil fuels:
    electrical energy from, 234
    hydrogen and, 104–106, 129
    Industrial Revolution and, 8
    methane and, 104–106
    national consumption, 12, 13, 234
    nitrous oxide and, 29–30
    recoverable reserves, 9
    (*See also* Coal)
Freon (*see* Chloroflourocarbons)
Freshwater, 27–29
Fuel cells, 86, 100–101, 277–280

Gasoline (*see* Petroleum)
Geothermal energy, 197–199, 213–222
Global energy demand, 10–13, 17–18, 234

Global warming, 3, 152–153
  carbon dioxide and, 24–26
  global trends, 22–37
  greenhouse effect, 70–71, 152, 325
  hydrogen and, 82–101
  methane and, 26–29
  nitrous oxide and, 29–31
  (See also Carbon dioxide;
    Chloroflourocarbons; Methane;
    Ozone)

Halocarbons (see Chlorofluorocarbons)
Heat pumps, 160–166
  (See also Chemical heat pump; Super
    Heat Pump;
Helium refrigerant, 1570
High Operating Temperature Electrolysis, 251
High Temperature Gas-cooled Reactor
    (HTGR), 3–4
  Bunsen reaction, 248
  components of, 236–241
  contrasted to other reactors, 235
  electrolysis of steam, 250–252
  fuel for, 237–238
  hydrogen production by, 3, 234–255
  iodine-sulfur cycle and, 248–250
  iron-halogen cycle and, 250
  methane reforming and, 241–247
  Mitsubishi, 268
  palladium membranes, 246–247
  process loops of, 243–245
  prototypes of, 254
  safety and, 236, 239–241
High Temperature Test Reactor, 254
Hydrocracking, 236
Hydrogen, 3, 109, 198
  air ratio, 123
  automobiles, 104–141
  backfire and, 121–126
  biological systems and, 88
  CO2 separation and, 55
  comparison studies, 7, 104–106
  compression of, 112–114
  electrolysis and, 62, 85–86, 250–252
  engine development, 121–134, 143
  gasoline engines and, 129
  High Pressure Tank (HP) method,
    109–121
  HTGR and, 3, 234–255
  hydrocarbon production, 84–85
  injection methods for, 121, 126–134

liquid hydrogen, 89–90, 99–100,
    109–121
metal hydrides, 88–89, 109–121
methane reforming process, 84
mixture formation, 141
photoelectrochemical production, 87–88
photosynthesis, 88
post-combustion losses, 137–141
pre-ignition, 106
safety and, 83, 126, 141–142
solar energy, 87–88
tank insulation, 114–118
thermochemical reactions, 62–85
throttle valve destruction, 126
transportation and storage, 88–99
UCLA SEED Rally, 123–126
waste utilization and, 87

Ice core samples, 24, 26–27
Industrial Revolution, 8–9
Italian systems, 310

Japan:
  acid rain in, 175–178, 195
  carbon dioxide levels, 15, 20
  chloroflourocarbons in, 168
  cogeneration in, 306
  energy consumption, 11, 328
  energy sources, 10, 264
  High Temperature Test Reactor, 254
  petroleum dependence, 14
  (See also Sunshine Project; Super Heat
    Pump)

Landfills, methane from, 27–29
Lawrence Livermore Laboratory, 218
Liquified Natural Gas (LNG) (see Methanol)
Lorenz cycle, 295

Magnesium, 43, 46
Marine environment, 52
Market allocation model, 198
Mark II experimental unit, 324
Methane, 26–29
  combustion energy for, 235
  fuel cells, 100
  methanol, 3, 62, 236, 256–257

343

Methane (*Cont.*):
    power plants, 51
    steam reforming, 84
    (*See also* Atmosphere; Global warming; Methanol)
Methylcyclohexane, 101
Molina-Rowland theory, 33
Montreal Protocol, 153–156, 167–169
Moonlight Project, 308

Natural gas (*see* Methanol)
New Zealand, 220
Nitrous oxide, 29–31, 179–182
    automobiles and, 61, 188–192
    boilers, 182–184
    carbon dioxide and, 58
    cogeneration and, 311
    diesel engines and, 187–188
    electric power and, 261–262
    electron irradiation and, 193, 195
    gas turbines and, 185–187
    internal engines, 184–188
    sources of, 27–28, 178–179
    SST and, 155
    stabilized levels of, 31
    Super Heat Pump, 284, 302–303
    (*See also* Acid rain; Nitrous Oxide)
Nuclear energy, 2, 83, 234–235
    (*See also* High Temperature Gas Cooled Reactor)

Oceans, 27–30, 43–47, 52, 198
Ozone, 3, 33–37, 145–152
    hydrogen and, 82
    levels of, 22–37
    Molina-Rowland theory, 33
    (*See also* Chloroflourocarbons; Global warming)

PAFT program, 156
Paper recycling, 21
Pentafluoropropylalcohol (5FP), 165
Petroleum (*see* Fossil fuels)
Photochemical smog, 155
Photoelectrochemical products, 87–88
Photosynthesis, 40–41, 88
Polystyrene, 163–164
Population increase, 2

Power plants, 18, 61–62
    (*See also* specific types)
Power transmission, 18
Public Utility Reguatory Act (PURPA), 310

Rankine cycle, 216
Refrigeration (*see* Chloroflourocarbons)
Rice paddies, 27–29

Safety:
    HTGR and, 236, 239–241
    hydrogen and, 83, 126, 141–142
    UCLA SEED Rally, 123–126
SEGS system, 206
Solar energy, 197–198, 200–210, 326
    $CO_2$ reduction and, 206
    future prospects of, 207
    hydrogen production by, 87–88
    photovoltaic system, 200–203
    solar electric generating system, 206
    STEPS and, 205–207
    system availability formula, 205
Solid state systems, 86
Soot, 63–68, 76
Southeast Asia, 10
Space technology, 99–100, 155
SST technology, 155
STEPS system, 205–207
Stirling engines, 311
Sulfur oxide, 192–195, 261–262
    (*See also* Acid rain)
Sunshine Project, 197, 215–217
Super Heat Pump, 155, 282–304
    chloroflourocarbons and, 299
    coefficient of performance, 282–284
    environmental protection and, 197–198, 284–285, 299–303
    load levelling, 284
    nitrous oxide and, 302–303
    waste heat utilization, 302–303

Thermal energy, 8–21
Thermodynamic laws, 325–326
Tidal energy, 198

UCLA SEED Rally, 123–126
United Kingdom, 310

United Nations Environmental Programme (UNEP), 1, 40, 155
Urban gas systems, 101
Urethane, 164
United States:
   cogeneration in, 310
   geothermal energy, 214, 216, 219
   HTGR prototype, 254
   Public Utility Regulatory Act (PURPA), 310

Vuilleumier cycle, 157

Waste heat, 302-303
Waste utilization, 27-29
West Germany, 254
Wetlands, 27-30
Wind energy, 198
Wood energy, 8-9